Bayesian Implementation

FUNDAMENTALS OF PURE AND APPLIED ECONOMICS

ADVISORY BOARD

Fundamentals of Pure and Applied Economics is an international series of titles divided by discipline into sections. A list of sections and their editors and of published titles may be found at the back of this volume.

Bayesian Implementation

Thomas R. Palfrey

California Institute of Technology, Pasadena, USA

and

Sanjay Srivastava

Carnegie Mellon University, Pittsburgh, Pennsylvania, USA

A volume in the Organization Theory
and Allocation Processes section
edited by
A. Postlewaite
University of Pennsylvania, Philadelphia, USA

Routledge
Taylor & Francis Group
LONDON AND NEW YORK

First published 1993 by Harwood Academic Publishers

2 Park Square, Milton Park, Abingdon, Oxfordshire OX14 4RN
52 Vanderbilt Avenue, New York, NY 10017

Routledge is an imprint of the Taylor & Francis Group, an informa business

First issued in paperback 2019

Library of Congress Cataloging-in-Publication Data

Palfrey, Thomas R., 1953-
  Bayesian implementation / Thomas R. Palfrey, Sanjay Srivastava ; a volume in the Organization theory and allocation processes section edited by A. Postlewaite.
    p. cm. — (Fundamentals of pure and applied economics ; v. 53)
  Includes bibliographical references and index.
  ISBN 3-7186-5314-1
  1. Economics, Mathematical. 2. Bayesian statistical decision theory. I. Srivastava, Sanjay, 1957- . II. Postlewaite, A.
III. Title. IV. Series.
HB135.P344 1993
330'.01'51—dc20   92-32494
                                                                      CIP

ISBN 13: 978-1-138-46949-5 (hbk)
ISBN 13: 978-3-718-65314-0 (pbk)

Contents

Introduction to the Series

Drawing on a personal network, an economist can still relatively easily stay well informed in the narrow field in which he works, but to keep up with the development of economics as a whole is a much more formidable challenge. Economists are confronted with difficulties associated with the rapid development of their discipline. There is a risk of 'balkanization' in economics, which may not be favorable to its development.

Fundamentals of Pure and Applied Economics has been created to meet this problem. The discipline of economics has been subdivided into sections (listed at the back of this volume). These sections comprise short books, each surveying the state of the art in a given area.

Each book starts with the basic elements and goes as far as the most advanced results. Each should be useful to professors needing material for lectures, to graduate students looking for a global view of a particular subject, to professional economists wishing to keep up with the development of their science, and to researchers seeking convenient information on questions that incidentally appear in their work.

Each book is thus a presentation of the state of the art in a particular field rather than a step-by-step analysis of the development of the literature. Each is a high-level presentation but accessible to anyone with a solid background in economics, whether engaged in business, government, international organizations, teaching, or research in related fields.

Three aspects of *Fundamentals of Pure and Applied Economics* should be emphasized:

— First, the project covers the whole field of economics, not only theoretical or mathematical economics.
— Second, the project is open-ended and the number of books is not

predetermined. If new and interesting areas appear, they will generate additional books.

— Last, all the books making up each section will later be grouped to constitute one or several volumes of an Encyclopedia of Economics.

The editors of the sections are outstanding economists who have selected as authors for the series some of the finest specialists in the world.

Bayesian Implementation

THOMAS R. PALFREY

California Institute of Technology, USA

SANJAY SRIVASTAVA

Carnegie Mellon University, USA

1. INTRODUCTION

The implementation problem lies at the heart of a theory of institutions. Simply stated, the goal of implementation theory is to investigate in a rigorous way the relationship between outcomes in a society and the institutions under which those outcomes arise. By outcomes, what is usually meant is the production and allocation of public and private goods and services. By institutions, what is meant is the set of rules according to which the allocation is decided upon and enforced (property rights, voting laws and voting rules, constitutions, contract law, etc.). In this monograph, we summarize and explain some recent results in a branch of implementation theory called Bayesian implementation.

Formally, an institution is represented by a mechanism. A mechanism is a complete description of the set of actions available to each agent in the economy and of the consequences of these actions. For example, in a bilateral bargaining situation, one such mechanism is the buyer's-bid double auction. The actions available to each agent are simply the real numbers: the buyer submits a bid, and the seller submits an offer. The consequence of a bid-offer pair is that trade takes place if and only if the buyer's bid is at least as high as the seller's offer, in which case the price equals the buyer's bid. There are many examples in public good situations, too. One simple public goods mechanism may ask each agent to contribute toward the cost of the public good, and the consequence of a set of contributions could be the level of

1

public good provision. In fact, most economic and political institutions can be viewed in principle as mechanisms.

Since we are interested in determining the outcomes of institutional arrangements, we also need a description of how individuals interact under the specific rules of a mechanism. This specification of individual behavior we call a behavioral rule, and it is constructed from the following basic elements. One element specifies the preferences of agents over the set of possible outcomes, since agents will typically need to compare the consequences of alternative actions. A second element specifies the information possessed by agents, since this information may bear directly on their ability to evaluate consequences. As a function of preferences and information, a behavioral rule defines how agents interact strategically with one another since the consequences of actions generally depend on the actions of all agents.

Several ways of modelling strategic interaction within a mechanism are provided by game theory, and the approach taken in Bayesian implementation is based on Bayesian game theory (Harsanyi [1967–68]). In this approach, the preferences and information of agents are modeled explicitly, beliefs are updated according to Bayes' rule, outcomes are evaluated according to expected utility theory, and finally, all agents are assumed to be Nash competitors, so that actions are chosen to maximize expected utility given the action choices (or strategies) of the other agents. Equilibrium is defined by a collection of strategies in which each agent is choosing a best strategy given the strategies of the other agents.

Given a mechanism and this specification of behavior, the equilibrium outcomes to a mechanism are simply the consequences of equilibrium strategies, and these outcomes represent the set of outcomes which can be achieved by the mechanism. Thus, we have arrived at a framework which allows for a precise study of outcomes achievable by institutions.

Viewed in this manner, the problem of characterizing the outcomes achievable by all mechanisms would seem formidable, since the set of possible mechanisms is extremely large. On the other hand, the set of possible outcomes is generally much smaller than the set of mechanisms. The problem is thus greatly simplified if we ask the following question: given a set of outcomes, is there a mechanism which achieves this set of outcomes? If there is, then the outcomes are said to be implementable.

Over the past decade, there has been considerable progress in characterizing the set of outcomes implementable in Bayesian equilibrium, and the purpose of this book is to summarize and explain these developments.

The relationship between institutions and outcomes also can be viewed from a 'design' standpoint, and in this guise is called the theory of mechanism design. The usual approach in mechanism design theory is to consider the point of view of a 'planner' or mechanism designer, who has some objective in mind and some obstacles to overcome and constraints to satisfy. The goal is to maximize some objective function that depends upon informatir n the planner does not have. That information is distributed among a set of individuals and part of the problem is how to collect it. This obstacle does not so much impose physical constraints, but rather takes the form of a collection of 'incentive compatibility constraints.' (Of course there are the more familiar feasibility constraints on the outcomes, as well.)

Ideally, a planner would simply read everyone's mind and impose the feasible oumome that maximizes his objective. But this is obviously unrealistic. At the very least, the planner needs to communicate with the individuals, through what has become known as a message space. This is often modelled as one-directional communication, where the individuals report messages to the planner, who then chooses a feasible outcome. Thus, in a nearly ideal world, the planner could simply ask individuals to report the relevant information, and everyone would do so in an honest way, after which the planner would simply impose the same outcome he would have selected by reading everyone's mind. This is 'nearly' ideal because the process of physically collecting such information uses up resources.

We are not even in the 'nearly ideal' world because an incentive problem arises: individuals may not be willing to honestly report this valuable information if they think reporting something else would lead to a decision that they prefer. These obstacles are overcome by judicious choice of a set of messages individuals are requested to report, together with an outcome rule to determine how these messages are used by the planner. This message space-outcome rule pair is precisely what we previously called a mechanism.

When formulated in this manner, the problem has the same basic structure as an engineering problem, since we are given an objective and a set of constraints. However, the problem is complicated by the fact that the mechanism itself is the choice variable. A major development in the theory of mechanism design is the revelation principle, which shows how the problem can be simplified by restricting attention to a specific and small class of mechanisms. This development allows the incentive constraints to be written as linear inequality constraints in the optimization problem of the planner. Thus, to the extent that game theory provides reasonably accurate models of behavior, achieving the planner's objective becomes a more or less standard constrained optimization problem.

Implementation theory studies the constraints that arise in this engineering problem. The basic ingredients in the mechanism design problem are:

— A set of agents, called a *society*
— A set of environments, called a *domain*
— A set of feasible decisions, called *outcomes*
— The objective of the planner, called a *social choice function*
— A way of collecting messages and making a decision, called a *mechanism*
— A behavioral rule, called an *equilibrium concept*

As noted, we adopt a fairly standard approach in mechanism design theory and consider behavioral rules that are derived from the principles of noncooperative game theory. These behavioral rules all have the property that individuals are assumed to respond to incentives.

The purpose of this monograph is to summarize and explain some recent results in a branch of implementation theory called Bayesian implementation. This limits our coverage of implementation theory in two important ways. First, we restrict attention to contributions to implementation theory that follow a Bayesian approach according to which players' probabilistic beliefs about their environment are modelled explicitly, and so the information structure is considered one aspect of the environment. Second, we look at the 'full' implementation problem. That is we want to design mechanisms that not only have equilibrium outcomes corresponding to the desired ones, but do not have equilibrium outcomes that are not desired ones.

This obviously leaves out much of the work in mechanism design theory. First of all, most of the recent work on Bayesian mechanism design addresses the first half of the full implementation problem, but not the second half. That is, the goal is to design a mechanism which has a desirable equilibrium outcome, but which may also have an undesirable equilibrium outcome. Second, in some work in mechanism design theory the players' priors about their environment really don't play an important role. In particular, the lines of research that investigate implementation in Nash equilibrium and implementation in dominant strategy equilibrium (i.e. strategy-proof mechanisms) do not include beliefs as an integral part of the environment. However, there is a close relationship between both of these lines of research and the literature we will be focussing on. First, the issue of incentive compatibility is central. Second, most of this work addresses the 'full' implementation problem.

The work on Nash implementation differs in that it assumes all players (except the mechanism designer) have complete information about the environment. Thus it is possible to treat many of the results that have been obtained in the Nash implementation literature as special cases (with degenerate priors) of more general results in the Bayesian implementation literature. At the end of Section 2 is a bibliographic guide to the literature on Nash implementation.

The work in dominant strategy implementation differs in the type of incentive compatibility that is required. The equilibrium concept of dominant strategies is much more restrictive, and col sequently far fewer social choice functions are dominant strategy incentive compatible than are Bayesian incentive compatible. It is so restrictive that when domains of environments are sufficiently broad, virtually no interesting social choice functions are implementable in dominant strategies. Thus one view of the Bayesian approach is that it is a retreat from the more appealing (or at least less controversial) dominant strategy approach — a retreat that is suggested by largely negative findings. We do not discuss dominant strategy implementation in this monograph. For a guide to the literature on dominant strategy implementation, the reader is referred to surveys by Muller and Satterthwaite [1985], Groves [1982], Laffont and Maskin [1982] and Dasgupta, Hammond, and Maskin [1979], and to the references in the more recent paper by Mookherjee and Reichelstein [1989].

We have organized the material in the following way. Section 2 sets out the basic notation and principles of equilibrium, efficiency and so forth. That section also introduces the two central obstacles to mechanism design: incentive compatibility and multiple equilibria. We explain and discuss the well-known revelation principle, which characterizes incentive compatible allocation rules. Finally that section presents two simple examples to illustrate how multiple equilibrium problems can undermine the revelation principle. Section 3 surveys the general results characterizing the set of implementable allocations in different kinds of domains under the behavioral rule implied by Bayesian Nash equilibrium. Section 4 looks at several specialized domains and specialized applications that are of particular interest to economics. These include classical pure exchange economies, the problem of eliciting preferences for public good provision, and bilateral bargaining. Section 5 addresses some of the problems that arise if a planner's ability to commit to a mechanism is limited, or if the individuals cannot be prevented from colluding against the planner by communicating with each other before playing the mechanism. Section 6 explores the implications of using other behavioral rules, including ones based on refinements of Bayesian Nash equilibrium. That final section also looks at weaker concepts of implementation, notably the idea of virtual implementation, whereby the planner is only concerned about finding a mechanism that approximately implements a social choice function.

2. A GENERAL MODEL

In this section, we present a basic model to study the Bayesian implementation problem. While the general formulation is somewhat abstract, many substantive problems of interest, such as public goods provision, auctions, and bargaining are special cases of the model, and these are addressed in subsequent sections.

A. The environment

In the basic model we consider, there is a finite set of agents, $\mathbb{I} = \{1, \ldots, I\}$. The set of feasible alternatives is A. Sometimes, we will consider random allocations; in this case, we will denote by $\mathbb{P}(A)$ the set of all probability distributions on A. In this case, the set of possible outcomes, or social choices, is $\mathbb{P}(A)$ rather than A. The exact structure of the set of feasible alternatives will, of course, vary as we consider different applications. For example, if we are in the setting of a pure exchange economy, A could be the set of all reallocations of some aggregate endowment.

Agents are distinguished by two things: their preferences and their information. Given that we are interested in analyzing problems of incomplete information, we need to formalize the notion that one agent may not know the preferences or information of any other agent. A common and convenient way to do this is to assume that a variable, called the *type* of an agent, summarizes the preferences and information of the agent. Let T^i denote the set of possible types for agent i, and assume that T^i is a metric space. Denote the set of all *type profiles* by T. For each i, denote by $T^{-i} = T^1 \times \ldots \times T^{i-1} \times T^{i+1} \times \ldots \times T^I$ the set of possible profiles of the types of all agents other than i.

We summarize i's information (or beliefs) about the other agents by means of a collection of conditional distribution functions on T^{-i}. For each t_i, we describe i's beliefs about the other agents by $G^i(\,\cdot\,|t_i)$. Note that we allow this distribution to change with i's type, which captures the notion that the types of an agent may correspond to different beliefs or information about the other agents. One special case is when $G^i(t_{-i}|t_i)$ is independent of t_i, which is widely referred to as the case of *independent types*. In that case, we will simply write $G^i(t_{-i})$. The general case is one of *dependent* or *correlated* types. We will sometimes consider what is called a *consistent beliefs* model (Harsanyi's [1967–68] terminology), in which all $G^i(\cdot)$ are 'consistent' with types being drawn from some common joint distribution, G, on T, so that $G^i(t_{-i}|t_i)$ is the conditional distribution of t_{-i} obtained from G given the drawn type, t_i.

Preferences are specified as follows. Given a profile of types t, and an alternative $a \in A$, the utility of agent i is given by $U^i(a, t)$. Note that the actual utility of an agent can depend on the types of all the others, even when these other types may be unknown to the agent. One special case is when $U^i(a, t) = U^i(a, t_i)$ for all a and t, which is the case of *private values*. We will refer to the more general case when preferences depend on the entire profile of types as *common values*.

The whole point of mechanism design is to select outcomes in a way that depends in a particular way on the information and preferences of the players. Thus, viewed from an *ex ante* point of view (i.e. before the types of the players are 'assigned'), the goal is not choosing an allocation *per se*, rather, it is choosing a type-contingent allocation (Harris and Townsend [1981]), much like the familiar state-contingent allocations in general equilibrium theory.

Such a type-contingent allocation is called an *allocation rule*, and it specifies an outcome in A (or $\mathbb{P}(A)$, if we are considering random allocation rules) for each possible vector of types $t \in T$. Allocation rules are required to be measurable functions, and we denote the set of all possible allocation rules by X. Given $x \in X$, $x(t)$ denotes the allocation in A (or $\mathbb{P}(A)$) specified by x for the type profile t.

An important concept in the theory of mechanism design with asymmetric information is that of *interim preferences*. Interim preferences are defined over allocation rules, not allocations *per se*. The term 'interim' denotes that point of time at which all agents have observed their own types, but do not know the types of the other agents. Agents typically take actions at this stage, and we therefore need to specify how they evaluate the consequences of their actions. Hence, we need to define their preferences at this interim stage, and this leads to the notion of interim preferences.[1]

Given an allocation rule, $x : T \to A$, the *interim utility* of x to type t_i of agent i is

$$V^i(x, t_i) = \int_{T^{-i}} U^i(x(t), t)\, dG^i(t_{-i} | t_i).$$

This definition of an interim utility function defines an interim preference relation on X for i that depends on t_i. If $V^i(x, t_i) > V^i(y, t_i)$, we can write $x P^i(t_i) y$. Similarly, we can define the weak interim preference relation, $R^i(t_i)$ and the interim indifference relation, $I^i(t_i)$.

Summarizing, we have the following basic building blocks:

1. Agents $\mathbb{I} = \{1, \ldots, I\}$
2. Feasible alternatives A ($\mathbb{P}(A)$)
3. Types $T = T^1 \times \ldots \times T^I$
4. Beliefs $\{G^i(t_{-i} | t_i),\ i \in \mathbb{I},\ t_i \in T^i\}$
5. Utility functions $\{U^i : T \times A \to \mathbb{R},\ i \in \mathbb{I}\}$
6. Allocation rules X
7. Interim preferences $\{V^i : T_i \times X \to \mathbb{R},\ i \in \mathbb{I}\}$

Notice that 1–5 above are the fundamental components of the mechanism design problem, and 6–7 are derived from 1–5. Thus, we

---

[1] Elsewhere, these have been called 'evaluation functions'. See Myerson (1985 p. 239) and the references cited there.

define an *environment*,[2] *e*, in terms of only the first 5 components. That is, $e = \langle \mathbb{I}, A, T, \{G^i\}, \{U^i\} \rangle$. A collection of environments, E, is called a *domain*.

B. Mechanisms and equilibrium

A *mechanism* is $\mu = (M, g)$, where $M = M^1 \times M^2 \times \ldots \times M^I$, and g is a function $g : M \to A$. The set M^i is the *message space* of agent i, and g is called the *outcome function*. For each $m \in M$, $g(m)$ yields an outcome in the set of alternatives.

Definition 2.1: A *strategy* for i is a function $\sigma^i : T^i \to M^i$.
Given a strategy profile $\sigma = (\sigma^1, \ldots, \sigma^I)$, the interim utility to i when of type t_i is given by

$$W^i(\sigma, t_i; M, g) = \int_{T^{-i}} U^i(g(\sigma(t), t) \, dG^i(t_{-i} | t_i)$$

Definition 2.2: σ is a *Bayesian equilibrium* if for all i and t_i,

$$W^i(\sigma, t_i; M, g) \geq W^i(\sigma^{-i}, \alpha^i, t_i; M, g) \text{ for all } \alpha^i : T^i \to M^i$$

If σ is an equilibrium to (M, g), then $g(\sigma)$ is called a Bayesian equilibrium outcome to (M, g). Note that $g(\sigma)$ is a function from T into A, and is thus an allocation rule. In much of what follows, we will refer to a Bayesian equilibrium of a mechanism simply as an equilibrium to the mechanism.[3]

C. The revelation principle and incentive compatibility

The central question in Bayesian incentive theory concerns the characterization of allocation rules which arise as equilibrium outcomes to a mechanism given individual preferences and information. On the face of it, it would seem very difficult to provide such a characterization. After all, the set of possible mechanisms is unimaginably large, and it can be quite difficult to characterize all possible equilibria

---

[2] Myerson [1985] calls this a 'Bayesian collective choice problem'.

[3] Note that we are restricting attention to pure strategies. Much of Bayesian implementation theory ignores mixed strategy equilibria. However, some of the results also hold for mixed strategy equilibria; we will indicate these later.

to all such mechanisms. However, it turns out that a result known as the *Revelation Principle* (Gibbard [1973], Myerson [1979], Harris and Townsend [1981]) makes this a feasible exercise. The principle says that in order to determine the set of allocation rules which arise as equilibrium outcomes to some mechanism, it suffices to focus on a particularly simple class of mechanisms and on a particularly simple type of equilibrium to these mechanisms. We turn to this next.

The revelation principle
Let (M, g) be a mechanism, and suppose that σ is an equilibrium to (M, g). By definition, we have

$$W^i(\sigma, t_i) \geq W^i(\sigma^{-i}, \alpha^i, t_i) \text{ for all } \alpha^i : T^i \to M^i,$$

and the equilibrium outcome under σ is $g(\sigma)$. Now, consider the 'direct' mechanism, (M_0, g_0), defined as follows:

$$M_0^i = T^i \text{ for all } i, \text{ and } g_0(t) = g(\sigma(t)).$$

This is called a direct mechanism since the message space of each agent is simply his set of types. Note that for any set of functions $\beta^1, \ldots, \beta^I$, with $\beta^i : T^i \to T^i$ for each i, we have

$$(*) \quad g(\sigma(\beta(t)) = g_0(\beta(t)) \text{ for all } t.$$

Then it must be true that the 'truthful' strategy $\sigma_0^i(t_i) = t_i$ is an equilibrium to (M_0, g_0). To see this, suppose this is not the case, so there is some i and a strategy for i, say $\beta^i : T^i \to T^i$, and some type for i, say t_i, such that

$$W^i(\sigma_0, t_i; M_0, g_0) < W^i(\sigma_0^{-i}, \beta^i, t_i; M_0, g_0).$$

This simply says that if all the other agents are using σ_0, then i does better at t_i by playing β^i than by playing σ_0^i. Now, consider the following strategy for i in the original mechanism (M, g):

$$\alpha^i(t_i) = \sigma^i(\beta^i(t_i)).$$

Since $\beta^i : T^i \to T^i$, α^i is a well defined strategy for i in the mechanism (M, g). Consider the outcomes in (M, g) when i uses α^i and all $j \neq i$ use σ^j. By (*), setting β^j to be the identity function for all $j \neq i$, we get

$$g(\sigma(\beta(t)) = g_0(\beta(t)) \text{ for all } t.$$

But then

$$W^i(\sigma_0, t_i; M_0, g_0) = W^i(\sigma^{-i}, \alpha^i, t_i; M, g), \text{ and}$$

$$W^i(\sigma_0, t_i; M_0, g_0) = W^i(\sigma^{-i}, \sigma^i, t_i; M, g),$$

which implies

$$W^i(\sigma, t_i; M, g) < W^i(\sigma^{-i}, \alpha^i, t_i; M, g),$$

which means that σ is not an equilibrium to (M, g), a contradiction. We conclude that σ_0 is an equilibrium to (M_0, g_0).

Hence, we have shown that if σ is an equilibrium to (M, g), then σ_0 is an equilibrium to (M_0, g_0). But since $g_0(t) = g(\sigma_0(t))$ for all t and $\sigma_0(t) = t$ for all t, we get $g_0(\sigma_0(t)) = g(\sigma(t))$ for all t. Thus the equilibrium outcome generated by σ_0 in the mechanism (M_0, g_0) is identical to that generated by σ in the mechanism (M, g).

Next, note that (M_0, g_0) is a particularly simple mechanism; each agent reports a type for himself. Further, σ_0^i is a particularly simple strategy; agent i reports his type truthfully. We have thus shown that if σ is an equilibrium strategy to (M, g), then $g(\sigma)$ can be reproduced by a mechanism in which the message space of each agent is his set of possible types and in which reporting your type truthfully is an equilibrium. This is precisely the revelation principle. Mechanisms in which $M^i = T^i$ for all i are called *direct mechanisms*. Direct mechanisms in which truthful reporting of types is an equilibrium are called *direct revelation mechanisms*.

Consider now the question we posed at the beginning of this section, namely: which allocation rules arise as equilibrium outcomes to a mechanism? Let $x: T \to A$ be an allocation rule which arises as the equilibrium to some mechanism, say (M, g), so that $g(\sigma) = x$ where σ is an equilibrium to (M, g). Then, we have established that x is the truthful equilibrium outcome of a direct mechanism, which we can denote by (T, g_0). But since the truthful strategy for each agent in (T, g_0) yields x as the outcome, we get that for each t, $g_0(t) = x(t)$. Thus, in the revelation game, we can set $x = g_0$.

Theorem 2.1 (The Revelation Principle): If x is an equilibrium outcome to some mechanism, then x is the truthful equilibrium outcome to the direct revelation mechanism (T, x).

Incentive compatibility

The revelation principle should not be interpreted as saying that we only expect revelation games to be played or that we should model all situations in terms of direct revelation mechanisms. It may well be the case that (M, g) is some intuitively pleasing or 'natural' mechanism while the direct mechanism has no obvious interpretation. What the principle does provide is a relatively straightforward way to determine whether it is ever *possible* for an allocation rule to be an equilibrium outcome. When this is true, we say that the allocation rule is *incentive compatible*.

It is now easy (in principle) to check if a particular x arises as an equilibrium outcome to some mechanism. We simply have to check if truth-telling is an equilibrium in the direct mechanism (T, x). We examine next exactly what this imposes on x.

In (T, x), consider the position of i when he is of type t_i, and all other agents are employing their truthful strategies. If i employs a strategy α^i, his expected payoff is

$$v^i(x; \alpha^i, t_i) = \int_{T^{-i}} U^i(x(t_{-i}, \alpha^i(t_i)), t) \, dG^i(t_{-i} | t_i).$$

For each $t_i \in T^i$, let $\alpha^i(t_i) = \tau_i \in T^i$. Then,

$$v^i(x; \alpha^i, t_i) = \int_{T^{-i}} U^i(x(t_{-i}, \tau_i), t) \, dG^i(t_{-i} | t_i),$$

which we can write simply as $v^i(x; \tau_i, t_i)$. Then, $v^i(x; \tau_i, t_i)$ is the expected payoff to i when all other agents are using their truthful strategies, i is of type t_i, and i reports τ_i. Thus, the requirement that truth be an equilibrium to (T, x), is summarized by:

Theorem 2.1': x is an equilibrium outcome to some mechanism if and only if for all i, for all $t_i \in T^i$, $V^i(x; t_i) \equiv v^i(x; t_i, t_i) \geq v^i(x; \tau_i, t_i)$, for all $\tau_i \in T^i$.

If x satisfies the conditions of this theorem, then x is said to be an *incentive compatible* allocation rule. The collection of inequality conditions in the theorem are called *incentive compatibility conditions*. Frequently, the content of the revelation principle is interpreted as simply requiring that x be incentive compatible. That is, given some environment e, if we want to see whether x can be made an equilibrium outcome in e by some appropriate choice of mechanism, then the problem reduces to one of verifying whether or not a particular collection of linear inequalities is satisfied.[4]

D. Efficiency

The power of the revelation principle stems from the fact that to determine whether an allocation rule is an equilibrium outcome to some mechanism, we simply have to check a system of linear inequalities. Frequently, we are interested in determining whether efficient allocations are incentive compatible. This raises the following question: what is an appropriate definition of efficiency when there is incomplete information?

The classical definition of efficiency is given by the Pareto criterion: an allocation is Pareto efficient if there is no other allocation which makes no one worse off while making some agents strictly better off. With incomplete information, two problems arise. First, when do we make this calculation? Do we do it before agents see their types (at the *ex ante* stage), after they see their types (at the *interim* stage), or do we check for efficiency given complete information about types (at the *ex post* stage)? Second, what happens if we propose an allocation rule which we believe to be efficient but which is *not* incentive compatible? In this case, the revelation principle tells us that there is *no* mechanism by which this allocation rule can be realized. Consequently, if we try to achieve this allocation rule, we cannot set up a mechanism to attain it, since it will be disadvantageous for some types of some agents to act as desired by the notion of efficiency.

The issues of efficiency with incomplete information received their first systematic treatment in Holmstrom and Myerson [1983]. They propose three definitions of efficiency with incomplete information, corresponding to the three points in time at which an allocation rule can be evaluated. These are called *ex ante, interim, and ex post incentive efficiency*, respectively. The qualifier 'incentive' is added because it is required that in any of the definitions, the allocation rule be incentive compatible. This distinguishes efficiency with incomplete information from the classical definition, which does not

---

[4] This point is evident in all the work on the revelation principle with incomplete information, going back at least to Harris and Townsend [1981] and Myerson [1979]. A particularly lucid explanation of the structure of these linear inequalities is in Ledyard [1986].

consider whether or not the allocation rule can be achieved by a mechanism.

Definition 2.3:   An allocation rule x: $T \to A$ is *ex ante incentive efficient* if it is incentive compatible, and there is no incentive compatible allocation rule y: $T \to A$ such that $\int u^i(y(t), t)dG^i(t) \geq \int u^i(x(t), t)dG^i(t)$ for all i, with strict inequality for some i.

Definition 2.4:   An allocation rule x: $T \to A$ is *interim incentive efficient* if it is incentive compatible and there is no incentive compatible allocation rule y: $T \to A$ such that $\int U^i(y(t), t)dG^i(t_{-i}|t_i) \geq \int U^i(x(t), t)dG^i(t_{-i}|t_i)$ for all i and t_i, with strict inequality holding for some i and t_i.

Definition 2.5:   An allocation rule x: $T \to A$ is *ex post incentive efficient* if it is incentive compatible and there is no incentive compatible allocation rule y: $T \to A$ such that $U^i(y(t), t) \geq U^i(x(t), t)$ for all i and t, with strict inequality for some i and t.

In what follows we will typically refer to an incentive efficient allocation rule simply as an efficient allocation rule. In all three definitions, it is not only important that the allocation rule x be incentive compatible, but it is equally important that it only be compared to other incentive compatible allocation rules.

It has been shown by Holmstrom and Myerson [1983] that *ex ante* (incentive) efficient allocation rules are also interim and *ex post* (incentive) efficient, while interim efficient allocation rules are also *ex post* (incentive) efficient. Further, an allocation rule x: $T \to A$ is interim efficient if and only if for every $t \in T$ it is not common knowledge that there exists an interim dominating allocation rule.

E. Implementation

There are some important reasons to be wary of overinterpreting the usefulness of the revelation principle. Most of these reasons are linked to the possibility of multiple equilibria. Suppose that (M, g) is a mechanism and x is an equilibrium outcome to (M, g). Then it is true that the direct mechanism (T, x) 'imitates' (M, g) in the sense that it reproduces one of its equilibria. However, this imitation may not be very precise if (M, g) has several equilibrium outcomes, some of which are not equilibria to (T, x); in this case, it might be necessary to construct a different direct mechanism for each equilibrium outcome of (M, g). These direct mechanisms would differ in that the outcome function (g_0 in the discussion above) would be different even though the message spaces would be the same.

Second, it may be true that the direct mechanism (T, x) has equilibrium outcomes which were not equilibrium outcomes in (M, g). Thus it is possible that x is the unique equilibrium outcome to some mechanism, while the direct mechanism (T, x) is plagued by multiple equilibria. We will present several examples demonstrating these points.

The implementation problem involves designing mechanisms to ensure that all equilibria have desirable properties.[5] The desirable properties are captured by a *Social Choice Correspondence* (SCC), F, which specifies, given an information structure and a set of alternatives, a set of desirable[6] allocation rules $\{x: T \to A\}$, so we have $F \subseteq X$. A *social choice function*, f, is a single-valued social choice correspondence, i.e. an allocation rule $f \in X$. We may want F to satisfy certain welfare properties, such as Pareto efficiency, responsiveness, monotonicity, etc. On the other hand, this approach also fits in naturally with the recent literature on principal-agent contract design and revenue maximizing auctions, both examples where dictatorial outcomes are being implemented. In general, however, we will not impose any restrictions on F. Later, we will explore the implications of a variety of specific efficiency criteria.

We say that a social choice function, f, is *weakly implementable*[7]

---

[5] Throughout, we assume that the feasible set is independent of the profile of types. Bayesian implementation with type-contingent feasibility is, at present, not well-understood.

[6] We use the loaded term 'desirable' intentionally. A social choice correspondence can be interpreted as a complete and precise description of the 'desirability standard' of the mechanism designer. Often in social choice theory and applications, we look at desirability standards that are implied by a few simple axioms, such as Pareto optimality or various fairness concepts, but in general social choice correspondences will not be so easily describable. These desirability standards are also called 'performance standards'. See, for example, Hurwicz [1972] or Mount and Reiter [1974].

[7] In view of Theorem 2.1, this is sometimes referred to as 'truthful implementability,' since f is weakly implementable if and only if f is the truthful Bayesian equilibrium of some direct mechanism. For more on this, see Repullo [1986] or Dasgupta, Hammond and Maskin [1979].

in e via Bayesian equilibrium if there exists some mechanism (M, g) and some Bayesian equilibrium σ of (M, g) in e such that $g(\sigma) = f$. The reason we call it 'weak' is that the implementing mechanism might have *another* equilibrium σ', with $\sigma' \neq \sigma$ and $g(\sigma') \neq g(\sigma)$. Thus if a planner intent on allocation rule x were to have the group play the mechanism (M, g) in e, the equilibrium outcome rule might not be x, as desired. Instead, an undesirable extraneous Bayesian equilibrium outcome $x' = g(\sigma')$ might occur. A stronger notion is that of *full implementation*, or simply *implementation*.

Definition 2.6:   A social choice function f is *implementable in e via Bayesian equilibrium* if there exists a mechanism (M, g) which has an equilibrium σ with $g(\sigma) = f$, and there does not exist an equilibrium σ' with $g(\sigma') \neq f$.

This definition of implementation extends in a natural way to social choice correspondences.[8]

Definition 2.7:   A social choice correspondence F is *implementable in e via Bayesian equilibrium* if there exists a mechanism (M, g) such that for each $x \in F$, there is an equilibrium, σ, with $g(\sigma) = x$, and there is no equilibrium σ' with $g(\sigma') \notin F$.

F. Two examples of multiple equilibria

The difference between weak implementation and implementation has to do with the possible existence of extraneous undesirable multiple equilibria in the implementing mechanism. We close this section with two very simple examples to help illustrate how this might arise:

Example 2.1
This is taken from Holmstrom and Myerson [1983], where it is used to illustrate a different point in the theory of mechanism design. There are two agents, who have private values and independent, equally likely, types. There are three feasible alternatives. Agent 1's prefer-

---

[8] Mookherjee and Reichelstein [1990a] use the term weakly implementable to describe a SCC for which there is a mechanism all of whose equilibria are contained in the SCC. Thus, in their definition, an SCC F is weakly implementable, if there exists a nonempty subset $F' \subseteq F$ that is implementable. So their definition of weak implementation of social choice *functions* is the same as the one given here.

	U_{11}	U_{12}	U_{21}	U_{22}
a	2	0	2	2
b	1	4	1	1
c	0	9	0	-8

FIGURE 2.1

ences are such that his first type prefers ranks a first and c last and his second type reverses these preferences. Both types of agent 2 prefer a over b over c, but 2's second type has a stronger dislike of c. The details are as follows (Figure 2.1):

$$I = \{1,2\}$$

$$T^1 = \{t_{11}, t_{12}\}$$

$$T^2 = \{t_{21}, t_{22}\}$$

$$A = \{a, b, c\}$$

$$G = \{\tfrac{1}{4}, \tfrac{1}{4}, \tfrac{1}{4}, \tfrac{1}{4}\}$$

The following allocation rule is *ex ante*, *interim*, and *ex post* (incentive) efficient and is weakly implementable by a direct mechanism (Figure 2.2):

In this mechanism, agent 2 can guarantee herself b by announcing t_{22} in the direct game. Agent 2 faces a 50/50 lottery between a and c by announcing t_{21} if player 1 announces honestly. Therefore, player 2 has no incentive to lie about her type if player 1 is telling the truth. Agent 1 chooses either a lottery between a and b (by announcing 't_{11}')

	t_{21}	t_{22}
t_{11}	a	b
t_{12}	c	b

FIGURE 2.2

or a lottery between c and b (by announcing 't_{12}'). Therefore, as long as player 2 announces honestly (or, in fact, announces 't_{21}' with any positive probability for either of the types) player 1 will have an incentive to announce honestly. Thus x is weakly implementable by this direct mechanism.

However, the direct mechanism has another equilibrium in which player 1 is always truthful and player 2 *always* announces 't_{22}'. This is sometimes referred to as a *pooling* equilibrium, since both types of player 2 adopt the same behavior.[9] This example illustrates that reliance on the revelation principle can be misleading if the stronger notion of implementation is required. In fact, for this example, our only hope of implementing x without extraneous equilibria requires looking to mechanisms with larger message spaces than the direct game. We will call such mechanisms either *augmented* or *indirect* mechanisms.[10]

An augmented mechanism must simultaneously accomplish three things. Goal 1: it must eliminate the extraneous equilibria. Goal 2: it must not add additional unwanted equilibria. Goal 3: it must leave the desired equilibrium outcome intact. Implementation theory, at its most abstract level, consists of characterizing exactly when such augmentations exist.

Returning to the example, the intuitive problem here is quite simple. One of the agents is lying. Therefore, a natural way to solve this is to give the *other* agent some extra messages which allow him to 'break' the extraneous equilibrium. In this simple example, this augmentation can be done with a single additional message component, which is used in conjunction with the reported type of player 1. We will call this message an 'objection', denoted ϕ. The message is interpreted as an announcement by player 1 that player 2 is misbehaving and that his reported messages should be interpreted to mean exactly the opposite of their literal meaning. That is, if player 2 reports 't_{21}', the mechanism should interpret it as meaning t_{22}, and vice versa. The augmented game is illustrated below in Figure 2.3.

---

[9] There is also an equilibrium where player 1 pools as well, always announcing 't_{12}'.

[10] Mookherjee and Reichelstein [1990a] refer to 'augmented direct mechanisms' where an augmented direct mechanism actually contains a component isomorphic to the direct mechanism which weakly implements f. It can be shown that implementation by augmented mechanism is equivalent to implementation by augmented direct mechanism.

$$t_{22} \quad t_{22}$$

	t_{22}	t_{22}
t_{11}	a	b
t_{12}	c	b
$t_{11} \cdot \phi$	b	a
$t_{12} \cdot \phi$	b	c

FIGURE 2.3

There is no longer an equilibrium with player 2 always announcing 't_{22}', since this would mean player 1's equilibrium behavior when he is a type t_{12} would be 't_{12}, ϕ'. But then player 2 would be better off announcing 't_{21}' when he is type t_{22}. Thus, the augmentation has accomplished goal #1: eliminate the unwanted equilibrium.

It is still an equilibrium for player 2 to announce truthfully and for player 1 to never object and always announce truthfully. Thus the augmentation satisfies goal #3: leave the desired equilibrium intact.

Finally, we must verify that goal #2 has been achieved: no new equilibria have been introduced into the augmented game. This is easily verified. Agent 2 has only four possible strategies: always tell the truth, always lie, always say 't_{21}', always say 't_{22}'. We have already shown that player 2 cannot always say 't_{22}' in equilibrium. He will not always say 't_{21}' in equilibrium either, since doing so would lead player 1 to choose 't_{12}' half the time (when player 1 actually is type t_{12}), and 't_{11}' the other half (when player 1 is type t_{11}). There is, however, an additional equilibrium, which is for player 2 to always lie and for player 1 to object and tell the truth. This equilibrium, however, produces exactly the same allocation rule as desired. Thus, the additional equilibrium is acceptable. This example illustrates that implementation does not mean unique implementation. There may be multiple equilibria (and generally *will* be if the desired SCC is multivalued), but none of the equilibria lie outside the SCC.

Example 2.2

We next present an example[11] in which an uninformed principal is trying to elicit information from two informed agents. The agents have preferences over feasible alternatives ('projects') that are directly opposed to those of principal. Incentive compatibility does not constrain the principal, who is designing the mechanism; i.e. the principal can design truth-telling mechanisms to weakly implement his most desired allocation rule. However, there is an undesirable equilibrium that creeps into the picture which is impossible to eliminate. Therefore, how can the principal solve this problem? The answer is: in the confines of a two-person direct game, it is impossible.

The principal must decide between two projects, *a* and *b*, and a third alternative which is to fire both agents. (Think of the agents as being on a fixed salary if they are not fired.) Each of the agents can be one of two possible types. Call the first agent γ, and his types γ_1 and γ_2. Similarly call the second agent β, and his types β_1 and β_2. The types are drawn independently, with prob $\{\gamma_1\}$ = prob $\{\beta_1\}$ = $q \neq 1/2$. The preferences of the two agents depend on the types of *both* agents, and the two agents have the same utility function. These preferences are displayed in Figures 2.4 and 2.5.

Thus agent γ knows whether *a* or *b* is the best project, and agent β knows very little, by himself. They each have a normalized von Neumann-Morgenstern utility function where the utility index of

Type profile				$U(\cdot)$
$\gamma_1\beta_1$	$\gamma_1\beta_2$	$\gamma_2\beta_1$	$\gamma_2\beta_2$	
a	a	b	b	1
b	c	c	a	v
c	b	a	c	0

FIGURE 2.4   Agents' utility function.

---

[11] A related example of this kind of multiple equilibrium problem can be found in Postlewaite and Schmeidler [1986 pp. 16–17].

their best outcome is always 1, the worst outcome is always 0, and the middle outcome is always v. Assume $v > \max \{q, 1 - q\}$.

The principal has exactly the opposite preferences between a and b. Furthermore, in the states $\gamma_1\beta_2$, and $\gamma_2\beta_1$, the principal would like to fire the agents. Thus the principal's utility function is:

Type profile				$U(\cdot)$
$\gamma_1\beta_1$	$\gamma_1\beta_2$	$\gamma_2\beta_1$	$\gamma_2\beta_2$	
b	c	c	a	1
c	b	a	c	v
a	a	b	b	0

FIGURE 2.5   The principal's utility function.

Therefore, the principal's first best allocation rule, x^*, is the one given in Figure 2.6.

It is easily verified that under the assumption $v > \max \{q, 1 - q\}$, x^* is incentive compatible. However, it is also easily verified that in the direct mechanism defined by x^*, there is also an equilibrium in which both agents always lie. Call this allocation x_L^*. There are no pooling equilibria.

It is also possible to show that the allocation x_L^* involves a joint strategy which cannot be eliminated even by using an indirect mechanism. That is, any mechanism which produces x^* as an equilibrium outcome must also produce x_L^* as an equilibrium outcome. While this is relatively tedious to try to show directly, the first characterization theorem in the next chapter shows that the question of whether or not an undesirable equilibrium can be 'selectively eliminated' can be answered by inspection of a set of linear inequalities, in much the same way that one can verify incentive compatibility.

Given the multiple equilibrium problem in this example, the planner must resort to more creative tactics if he is to prevent the agents from 'colluding' on the undesirable equilibrium. There are several ways this can be done. One thing that works is to simply hire an uninformed party with similar preferences to his, or perhaps simpler, become the

FIGURE 2.6   The principal's first best allocation rule, x^*

third player in the game himself. The key to understanding when this will work lies in a subtle feature of (Bayesian) Nash equilibrium, namely that the equilibrium strategies of all the agents are common knowledge. Thus, we simply give the principal a strategy that he can use to advantage if the informed agents are using a joint strategy that produces the 'bad' outcome, x_L^*. It turns out that we do it in a similar way as the previous example.

If the uninformed principal *himself* joins the game and he gives himself a non-trivial message space[12], the problem can be overcome. In fact this is a general principle in full implementation theory with incomplete information. Often much can be gained by either including an extra player with *no information* (see Palfrey [1990]), or, for similar reasons, eliciting messages from the agents before they learn their types (Matsushima, 1990a).

In the case of this example, the principal's message space is {'Truth', 'Lie'}, the informed players use their 'direct' message space, and the outcome function is shown in Figure 2.7.

In other words, if the principal says 'Truth' the game is the same as the original direct game. If the principal says 'Lie' the interpretation of the agents' messages is reversed. Again, as before, this simply *converts* the bad equilibrium into a good equilibrium rather than actually eliminating the bad equilibrium. There are no pooling equilibria. Thus the new 3-person game has 2 equilibria, both of which produce the desired outcome.

---

[12] Or, alternatively, the principal hires an agent with the same preferences as himself.

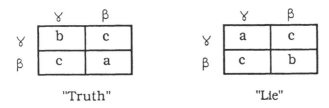

"Truth"   "Lie"

FIGURE 2.7

3. CHARACTERIZING IMPLEMENTABLE ALLOCATION RULES

In this section, we analyze the extent to which the multiplicity problems discussed in Section 2 can be avoided by judicious construction of a mechanism. We are interested in several questions. First, we wish to know which allocation rules can be made *unique* outcomes to a mechanism. Second, we would like to get some idea of which allocation rules cannot be made unique equilibrium outcomes. Third, we are interested in collections of allocation rules, or social choice correspondences, which are implementable.

Let x: $T \to A$ be an allocation rule and suppose we wish to know if x can be made the unique equilibrium outcome to a mechanism. We already know that x must satisfy the incentive compatibility constraints developed in Section 2, since these constraints are necessary for x to be an equilibrium outcome to any mechanism. Hence we already know that x is the truthful equilibrium outcome to the direct mechanism (T, x). We start by asking whether this direct mechanism has any other equilibrium outcomes.

A. Diffuse information structures

In analyzing this question, we will first make an assumption about the information structure to simplify the presentation of the most basic material. Specifically, we assume that information is *diffuse* in the sense that no strict subset of agents can pool their information and rule out certain types of the other agents. This is easiest to understand in the context of an 'objective' Bayesian model in which types are drawn from a commonly known probability distribution G on a finite set of types, T. In such a model, diffuseness corresponds to $G(t) > 0$ for all t. For all i and t,

$$G^i(t_{-i}|t_i) = \frac{G(t)}{\displaystyle\sum_{\tau_{-i} \in T^{-i}} G(\tau_{-i}, t_i)}$$

Consider the set of strategies available to agent i in the direct mechanism (T, x). Recall that a strategy for i in the direct mechanism is a function $\alpha^i : T^i \to T^i$, so that the set of available strategies for i is the set of all functions from T^i into T^i, the identity function being the *truthful strategy*. We call α^i a *deceptive strategy* or, more simply, a *deception* by i. The reason behind this terminology is straightforward; if $\alpha^i(t_i) \neq t_i$, then when i is of type t_i, i falsely reports the type $\alpha^i(t_i)$, which can be interpreted as an attempt to deceive other agents about i's true type. If α^i is the truthful strategy, then we call it the identity deception and denote it by α_0^i. A collection of deceptions $\alpha = (\alpha^1, \ldots, \alpha^I)$ is called a *joint deception*, and the set of all joint deceptions is denoted by $\mathbb{A}$.

Incentive compatibility already implies that if every agent other than i is using his truthful strategy, then no deceptive strategy gives i a higher expected payoff than the truthful strategy; this is the content of the incentive compatibility conditions developed in Section 2. The uniqueness question in the direct mechanism thus reduces to asking whether there is a set of joint deceptions which form an equilibrium.

Suppose $\alpha = (\alpha^1, \ldots, \alpha^I)$ is a joint deception which is an equilibrium to the direct mechanism. Then, when the realized vector of types is t, the outcome is $x_\alpha(t) \equiv x(\alpha(t)) = x(\alpha^1(t_1), \ldots, \alpha^I(t_I))$. Clearly, the problem arises if, for some t, $x_\alpha(t) \neq x(t)$.

Unfortunately, as shown by the examples in the previous section, it is easy to find examples of direct mechanisms which have multiple equilibria of this sort. Once we take seriously the possibility that direct mechanisms can have undesirable outcomes, we are naturally led to inquire into ways in which these possibilities can be avoided. There are at least three ways in which we might proceed. The first is to abandon direct mechanisms. After all, the revelation principle does not say that we expect economic institutions to be direct mechanisms. Given that multiple equilibria create coordination problems, one might expect economic institutions to develop practical ways to avoid exactly these kinds of problems. Second, we could try to argue that if faced with a direct mechanism, it makes intuitive sense to expect agents to play their truthful strategy; after all, it achieves the desired outcome and is a 'natural' equilibrium. As the examples show, it is frequently the case that playing a non-truthful strategy can lead to better outcomes for some or all agents, which undermines this argument. Third, we could attempt to examine a different notion of equilibrium. In this case, the argument is that Bayesian equilibrium does not place sufficient restrictions on 'reasonable' behavior, and some of the undesirable Bayesian equilibria would be unlikely to be played because they require players to behave implausibly (or to expect others to behave implausibly).

Implementation theory investigates a combination of the first and third directions. In this section, we will show that by moving away from direct mechanisms, it is possible to resolve the multiplicity problem in many problems of interest. In a later section, we show that if we combine the power of indirect mechanisms with slightly stronger equilibrium requirements, then in many more cases we can completely eliminate the multiplicity problem.

Let $\alpha = (\alpha^1, \ldots, \alpha^I)$, $\alpha^{-i} = (\alpha^1, \alpha^2, \ldots, \alpha^{i-1}, \alpha^{i+1}, \ldots, \alpha^I)$ so that $\alpha = (\alpha^{-i}, \alpha^i)$. As we have argued, every candidate for equilibrium in the direct game is a (joint) deception α. If α is being used, the outcome to the direct game is x_α, where $x_\alpha(t) = x(\alpha(t))$ for all t. Suppose that α is an equilibrium and that $x_\alpha \neq x$, so we have an undesired outcome to the direct game.

The question being posed can now be rephrased as asking when it is possible to add strategies to the direct game so that $x_\alpha \neq x$ is not an equilibrium outcome. This method is called *selective elimination* by Mookherjee and Reichelstein [1990a], and underlies most of the techniques which have been developed to eliminate unwanted equilibria.

Consider giving agent i a single additional message, say m^i. For each t, define y by: $y(t_{-i}, \alpha^i(t_i)) = g(t_{-i}, m^i)$, the outcome when i plays m^i and the other agents play t_{-i}.[13] To ensure that α is not an equilibrium, we must be able to find some such y so that m^i is strictly better than α^i for some type t_i for some agent i when the other agents are playing α^{-i}. If i plays m^i and the others use α^{-i}, the outcome at

---

[13] Observe that y depends only on t_{-i} (it is constant in t_i). Therefore for any α^i, and for any r_i, $y(t_{-i}, t_i) = y(t_{-i}, \alpha^i(t_i))$.

t is $y(\alpha^{-i}(t_{-i}), \alpha^i(t_i))$. If i uses α^i and the others use α^{-i}, the outcome at t is $x(\alpha^{-i}(t_{-i}), \alpha^i(t_i))$. Thus, x_α cannot be prevented from being an equilibrium outcome unless there exists some i, some t_i, and some $y: T^{-i} \to A$ such that $V^i(x_\alpha, t_i) < V^i(y_\alpha, t_i)$.

If there exists i, t_i, and y such that the above inequality is satisfied, then α cannot be an equilibrium in the new mechanism that has been augmented to include m^i. However, we must also be careful that introducing m^i in this way does not prevent x from being an equilibrium outcome, i.e. we still want truth telling to be an equilibrium. Thus, we must also have that *for all $t_i' \in T^i$, $V^i(x, t_i') \geq V^i(y, t_i')$*. If, whenever $x_\alpha \neq x$, there exists i, t_i, and y that satisfy the above inequality conditions, then x is called *Bayesian monotonic*.

Definition 3.1: $x: T \to A$ satisfies *Bayesian monotonicity* if for any deception α such that $x_\alpha(t) \neq x(t)$ for some t, there exist i, t_i and an allocation rule $y: T \to A$ such that $V^i(x_\alpha, t_i) < V^i(y_\alpha, t_i)$, and, for all t_i', $V^i(x, t_i') \geq V^i(y, t_i')$.

We can now summarize the two key necessary conditions for Bayesian implementation in the following theorem.

Theorem 3.1: If $x: T \to A$ is implementable, then

(i) x satisfies Bayesian monotonicity,
(ii) x is incentive compatible.

The next question concerns sufficiency. We provide a sufficiency proof in a particularly simple setting in which the following assumption is satisfied. We assume that each agent i has a best element, say $b^i \in A$, which is independent of his type. Further, assume that b^i is strictly preferred to b^j for all $i \neq j$. Note that this is always satisfied in a pure exchange economy (let b^i be the aggregate endowment). This assumption allows us to illustrate the intuition behind the proof in a simple manner. We indicate below the more general sufficiency results which have been obtained.

Theorem 3.2: Let x be an incentive compatible allocation rule which satisfies Bayesian monotonicity. If the above assumption is satisfied, T is finite, $I > 2$, and information is diffuse, then x is implementable.

Proof: The proof is by construction of an implementing mechanism. Let $M^i = T^i \times \{X^{-i} \cup \{0, 1, 2, 3, \ldots\}\}$. That is, each player reports his type and either an allocation rule that is constant in his report, $y: T^{-i} \to A$, or a nonnegative integer. We call the first part of the message space the 'direct' part, and the second part the 'indirect' part. The outcome function is defined by the following rules.

1. If $m^i = (t_i, 0)$ for all i, then

$$g(m) = x(t).$$

2. If there exists i such that $m^i = (t_i, n), n > 0$, or $m^i = (t_i, y)$, $y \in X^{-i}$, and $m^j = (t_j, 0)$ for all $j \neq i$, then

$$g(m) = x(t) \text{ if } m^i = (t_i, n) \text{ or if } V^i(x, t_i') < V^i(y, t_i') \text{ for some } t_i' \in T^i$$

$$= y(t) \text{ if } V^i(x, t_i') \geq V^i(y, t_i') \text{ for all } t_i' \in T^i.$$

3. For any other joint message, $g(m) = b^k$, where k is the agent reporting the highest number in the second component of his message, and ties are broken in favor of the lowest j who reports the highest number. A report of an allocation rule counts as 1 in this calculation.

Thus, we see that the outcome function is defined by dividing the message space into three regions. We call the first region the 'agreement' region. Here, everyone simply reports a type and a '0'.

The second region is the 'unilateral deviation' region. Here, exactly one agent, i, reports something other than 0 in the indirect part of his message. This deviation is ignored (i.e. treated like a report of 0) with one exception. If i reports $y \in X^{-i}$, the outcome is $y(t)$ if $V^i(x, t_i') \geq V^i(y, t_i')$ for all $t_i' \in T^i$. Interpret the report of y as a request by i to substitute the allocation rule y for x. Obviously, the mechanism cannot honor all requests of this sort, since then x would typically not be an equilibrium outcome. The mechanism only grants such requests when the substitution makes i worse off for every possible type he could be, *assuming the others are reporting truthfully.*

The third region is the 'disagreement' region, and consists of messages in which at least two agents report something other than 0 in the indirect part of their message. In this region, the agent reporting the highest integer can obtain his best allocation.

Now, observe that σ, given by $\sigma^i(t_i) = (t_i, 0)$ for all i, is a

Bayesian equilibrium. This follows since x is incentive compatible, and any unilateral deviation by any agent can only result in receiving some y, which is no better than x given that all other agents are reporting truthfully.

Second, observe that all equilibria must take the form $\sigma^i(t_i) = \{\alpha^i(t_i), 0\}$. This is due to the assumption of a finite number of types[14] and the assumption that all best elements are distinct. If some agents are reporting positive numbers or allocations at some types, some other agent can always report a number which is greater than anyone else's at any of their types and get his best element. There is no equilibrium with positive numbers being reported since these best elements are distinct.

Third, consider any $\alpha = \{\alpha^1, \ldots, \alpha^I\}$ such that $x_\alpha \neq x$. Then, by Bayesian monotonicity, there is an agent, say i, and a type of this agent, say t_i, and an allocation y such that at t_i, i prefers y to x given that the others are using the α strategies, but for all t_i', x is interim preferred to y. Therefore, if i reports $(\alpha^i(t_i), y)$, then the outcome is y_α instead of x_α, which makes i better off. Hence such an α is not an equilibrium.

Thus, all equilibria take the form $\sigma^i(t_i) = (\alpha^i(t_i), 0)$ with $x_\alpha = x$, so every equilibrium outcome to the mechanism is x. This concludes the proof. ■ ■ ■

The proof of this theorem illustrates the main features of sufficiency arguments in the Bayesian implementation literature. In what follows, we discuss alternative characterizations which have been obtained in more specific settings and also in more general settings.

The first alternative we examine is the case of a transferable money good. Most of the examples in the previous sections had a discrete allocation space. In many applications of interest, however, there exists a continuous good which everyone values positively, and this good exists in fixed supply in the economy. We will call such environments 'trading environments'. The most common example of such an environment is a pure exchange economy. In the contracting literature dealing with agency problems and bilateral trade with moral hazard and adverse selection, a simpler kind of scarcity environment is often

---

[14] The assumption of finite types can be dispensed with, but this introduces some technicalities that make the argument less transparent.

used: the case where there is a 'transferable utility good' which we will simply call the money good. This good enters into individual utility functions as a linear additive term, and such utility functions are often referred to as quasi-linear. In addition, it is often assumed that the utility functions are defined this way for any positive or negative allocation of money. Moreover, preferences over the non-money part of the allocation space does not depend on the money allocation.

Formally, let D be the set of feasible (non-money) allocations in the environment, and assume that the fixed supply of the money good equals 0. Then the set of feasible allocations in such environments is $A = D \times \mathbb{R}^I_o$, where

$$\mathbb{R}^I_o = \left\{ w \in \mathbb{R}^I \mid \sum_{i=1}^{I} w_i = 0 \right\}$$

and the (type-dependent) utility functions are given by:

$$U_i(a, w, t) = V_i(a, t) + w_i, \quad \text{for } a \in D, \, w \in \mathbb{R}^I_o.$$

The ability to freely transfer utility via sidepayments of the money good provides a mechanism designer with a lot of flexibility for 'breaking' equilibria by rewarding agents who report suspected deceptions of the other agents. A simplifying feature of these models is that individuals value the money good in a way that is known to the planner; the marginal utility of the money good does not depend on the D-allocation, nor does it depend on the type of the agent in question or the types of the other agents.

The power of this 'independence' property of individual preferences for the money good can be seen by referring back to the definition of Bayesian monotonicity earlier in this section. Recall that if we want x to be an equilibrium allocation rule, but we do not want x_α to be an equilibrium allocation rule, monotonicity requires the existence, for some type of individual i, of an allocation rule y that must satisfy some linear inequalities. In particular, if everyone else is telling the truth, then y must make all types of agent i interim worse off than x.

The money good allows the use of a very special such y. In particular, often it is possible to find such a y which simply equals x plus an additional balanced transfer of the money good, where that transfer depends only on the reported types of the agents other than i.

We first illustrate this in Example 2.2 of the previous section. Recall that in this example, there are two agents and a principal. The principal must decide between two projects, a and b, and a third alternative which is to fire both agents. The agents have some differential information about which project is better for the principal, but both prefer to work on the worst project.

As argued in Section 2, x^* is incentive compatible, but there is also an equilibrium in the direct mechanism in which both agents always lie, producing allocation x_L^*, which cannot be eliminated by using an indirect mechanism. That is, any mechanism which produces x^* as an equilibrium outcome must also produce x_L^* as an equilibrium outcome.[15]

However, one can show that if there is a 'transferable utility good', call it money, then there will exist an indirect mechanism to implement x^* uniquely. The basic intuition follows closely a result in Palfrey [1990], and relies heavily on three features of the environment:

(1) Side payments are possible.
(2) Player types are independently distributed.
(3) A 'consistency condition' is satisfied.

This consistency condition, which is used in Matsushima [1990a], requires simply that in any direct game, truthful reporting is the *only* strategy an agent has which produces a probability distribution of his reported types that is exactly the same as (i.e. *consistent with*) the distribution of his type that the other agents believe. For any i, and $j \neq i$, let $G^j(t_i)$ be the beliefs that agent j has about agent i's type. Let $G_\alpha^j(t_i)$ be the j-believed probability distribution of reported types by agent i under the deception α. Then *consistency* says $G_\alpha^j(t_i) = G^j(t_i)$ if and only if α is the truthful strategy.

If we assume priors are 'objective', then this reduces to a requirement that the distribution of i's reported type is the same as the *true* distribution of his type if and only if he reports truthfully. We will say that such environments satisfy the condition of *No Consistent Deceptions* (*NCD*). In such environments, often we can eliminate equilibria using extremely simple and natural mechanisms, by specifying small (balanced) sidepayments, or fines and rewards, that guarantee the unique implementation of an incentive compatible allocation rule.

---

[15] The reader might find it a useful exercise to verify this fact directly, by checking the inequalities in the definition of Bayesian monotonicity.

Before stating a general result, first consider the example. The condition of NCD is satisfied since $q \neq .5$. Suppose both agents are using the strategy of always lying. Then either agent, say agent γ, is reporting γ_1 with probability $1 - q$ and γ_2 with probability q. With truthtelling, these probabilities are reversed. Without loss of generality, suppose $q > 1/2$. Now consider a transfer rule in addition to the allocation rule x^*, where the rule calls for γ to pay β \$1 if β reports β_1, and for β to pay γ $\left(\dfrac{q}{1 - q} - \varepsilon \right)$ if β reports β_2.[16] Such an 'augmented' allocation has the property that, for small ε, it makes γ better off than x_L^* if β is always lying, but it makes γ worse off than x^* if β is telling the truth. Independence implies that the above claim is true regardless of γ's true type, since the transfer is constructed independently of his type. Thus, we can augment the direct mechanism in the following way to eliminate undesirable equilibria.

Let $\tau = \{ b : \{\beta_1, \beta_2\} \rightarrow \mathbb{R} \}$, so τ is the set of transfers from γ to β as a function of a reported type by β. Let $\bar{\tau} = \{ y \in \tau$ such that $q + y(\beta_1) + (1 - q)y(\beta_2) > 0 \} \cup \{0\}$. Let the message space of β be $\{\beta_1, \beta_2\}$ as in direct mechanism, but let the message space of γ be $\{\gamma_1, \gamma_2\} \times \bar{\tau}$. That is, γ reports his type together with a transfer rule which must either be 0 or must make γ worse off when β is always reporting truthfully. The planner then allocates $x(\hat{t}_\gamma, \hat{t}_\beta)$, and performs the requested transfer rule by γ, call it $\hat{t}(\hat{y}_\beta)$.

The unique equilibrium of the mechanism is for both agents to report truthfully and for γ to request 0 transfers. That such constructions are generally available with independent types is proven next.

Theorem 3.3:   If types are independent, NCD is satisfied, and there is a money good, then x is incentive compatible if and only if x is fully implementable.

Proof: (if): This is trivial since incentive compatibility is necessary for implementation.

---

[16] If $q < \frac{1}{2}$, then the transfers are reversed, so γ receives $1 - \varepsilon$ from β when β_1 is reported and pays $\dfrac{q}{1 - q} \neq \beta$ when β_2 is reported.

(only if): Consider the following mechanism. For each i, define

$$W_i = \left\{ w: T^{-i} \to \mathbb{R}^I \mid \sum_{j=1}^{I} w_j(t_{-i}) = 0 \, \forall \, t_{-i} \in T^{-i} \text{ and} \right.$$

$$\left. \sum_{t_{-i}} G_i(t_{-i}) w_i(t_{-i}) < 0 \right\}$$

The message spaces are:

$$M^i = T^i \times (W_i \cup 0)$$

$$g(m) = \begin{cases} x(t) & \text{if } m^i = (t_i, 0) \quad \forall i \\ x(t) + w_i(t_{-i}) & \text{otherwise,} \\ & \text{where } i \text{ is the agent with the lowest} \\ & \text{index, who is not reporting a 0} \end{cases}$$

Since x is incentive compatible, if everyone other than i reports truthfully, then i's best response is to report truthfully. As long as everyone else reports 0 as well as reporting truthfully, then i's *strict* best response is also to report 0. Therefore, it is an equilibrium for all agents to report truthfully and to report zero. It is not an equilibrium for all agents to be reporting truthfully and for some agents to not report zero because the i determining $w_i(\cdot)$ is better off reporting $k w_i(\cdot)$ where $0 < k < 1$.

Thus we must simply show that there can be no equilibrium with a deception. Let α be a deception. Since α is not consistent with truthful reporting, it cannot be an equilibrium with everyone reporting $(\alpha_i(t_i), 0)$ since for some i there exists a $w_i(\cdot) \in W_i$ such that

$$\sum_{t_{-i}} G^i(t_{-i}) w_i(t_{-i}) < 0 \text{ but } \sum_{t_{-i}} G^i(\alpha(t_{-i}) w_i(t_{-i})) > 0$$

and such that i is better off reporting $(\alpha(t_i), w(\cdot))$ instead of $(\alpha(t_i), 0)$. Next suppose that some subset of the individuals report $w_i(\cdot) \neq 0$. Then $w_i(\cdot)$ is the one reported by the agent with the lowest index, call it i. Then one of the following holds:

$$\sum_{t_{-i}} G^i(\alpha(t_{-i})) w_i(t_{-i}) > 0$$

$$\sum_{t_{-i}} G^i(\alpha(t_{-i})) w_i(t_{-i}) = 0$$

$$\sum_{t_{-i}} G^i(\alpha(t_{-i})) \, w_i(t_{-i}) < 0$$

In the first case, i improves his payoff by reporting $kw_i(\cdot)$ where $k > 1$. In the third case, i improves his payoff by reporting $kw_i(\cdot)$ where $k < 1$. In the second case, we use the fact that $\Sigma_{t_{-i}} G_i(t_{-i}) \, w_i(t_{-i}) = 0$.

The NCD condition then implies that there exists $\hat{w}(t_{-i})$

$$\sum_{t_{-i}} G^i(t_{-i})) \, \hat{w}_i(t_{-i}) < 0$$

$$\sum_{t_{-i}} G^i(\alpha(t_{-i})) \, \hat{w}_i(t_{-i}) > 0.$$

Thus i can improve by reporting $\hat{w}(\cdot)$ instead of $w_i(\cdot)$. ■ ■ ■

Several remarks are in order about this result. First, even though we require independent types there are clearly a large variety of cases in which this is not needed. The most obvious case involves the selective elimination of equilibria with pooling of types. If there is pooling then some types of some agents are *never* reported in equilibrium (assuming a finite set of types). When this is the case, and if $G^i(t_{-i}|t_i)$ always has full support, such pooling equilibria can nearly always be eliminated. To see this let type t_j of agent j be a type that is never reported in the pooling equilibrium and let i be any agent other than agent j. Simply augment agent j's strategy space in a similar manner to the construction in Theorem 3.3. Now i can break the pooling equilibrium by requesting a side payment scheme in which i must pay a very large transfer if j reports t_j, and receives a very large transfer if j does not report t_j. This type of construction does not require anything nearly as strong as independence. What is needed is a very mild condition on the consistency of players' priors in the form of a common support assumption. The construction could only fail if $G^i(t_j|t_i) = 0$ for all $i \neq j$ and for all $t_i \in T^i$.

Second, for obvious reasons, the requirement of no consistent deceptions could be relaxed to a requirement of no consistent *equilibrium* deceptions, but this may be a difficult thing to check. It would essentially reduce to proving that there are no separating equilibria in the direct game other than the truthful equilibrium. Third, independence can be dispensed with entirely if there is an uninformed agent (see Palfrey [1990]). To the extent that the planner could make himself an agent or could pay a third party (an 'auditor') to play the part of the uninformed agent, this is a very strong result. Fourth, this result applies to continuous type models, but in a way that is essentially vacuous since the assumption of no consistent deceptions will generally fail. Fifth, the existence of a money good that enters separately and linearly can be relaxed. The linearity assumption that agents are all risk neutral with respect to money can be relaxed quite a bit, but preferences for money must still be type independent. Therefore, separability of preferences for money and preferences over the rest of the allocation space would seem to be more difficult to relax.

The next characterization we discuss relaxes both the assumption of independent types and the assumption that there is transferable utility. Both Matsushima [1990b] and Jackson [1991] obtain results for this case, by assuming that there is some degree of conflict between the agents. We present a version of the characterization and proof of Matsushima, in part because his technique of proof is innovative and illustrative of the general technique of 'selective elimination' that is used in other constructive proofs, an idea found in Mookherjee and Reichelstein [1990a].

The basic assumptions Matsushima begins with are that there are a finite number of types and that information is diffuse. Formally, he assumes that

$$G^i(t_{-i}|t_i) > 0 \quad \text{for all } i \text{ and } t$$

The next assumption is that an agent's most preferred allocation depends only on that agent's type. We call this assumption *best-element private-values* (BEPV), because it restricts the degree to which agents' preferences may depend on other agents' types, and this restriction only affects best elements. This assumption also contains the implied assumption that every type of every agent has a 'best element'.[17] The assumption of the existence of best elements appears elsewhere in the implementation literature (see, for example, Palfrey and Srivastava [1989b] and Jackson, Palfrey and Srivastava [1992]).

---

[17] Matsushima assumes that the allocation space A is compact and that utility functions are continuous, which guarantees the existence of best elements.

Assumption (BEPV): For every $i \in I, t_i \in T^i$ there exists $b^i(t_i) \in A$ such that $U^i(b^i(t_i), t) \geq U^i(a, t) \, \forall \, a \in A, \forall \, t_{-i} \in T_{-i}$.

A second assumption is that agents disagree about best elements. To simplify the statement of this next assumption, we add the assumption that best elements are uniquely defined.[18]

Assumption SCP (Strongly Conflicting Preferences): For all $a \in A$, $t \in T, \#\{i \,|\, b^i(t_i) = a\} \leq I - 2$.

This assumption is closely related to an assumption called *No Veto Power* that is often assumed in the Nash implementation literature and which is always satisfied in environments with strongly conflicting preferences.[19]

Theorem 3.4 (Matsushima, Jackson): If $I \geq 3$, information is diffuse, best elements exist and are unique, and Assumptions BEPV and SCP are satisfied, then a social choice correspondence F is fully implementable in Bayesian equilibrium if and only if it is Bayesian monotonic and every $x \in F$ is incentive compatible.

Proof: Earlier arguments establish the *only if* part of this theorem. Matsushima establishes the *if* part by constructing a mechanism in roughly the following way.

For every $x \in F$ and for every deception α, either $x_\alpha \in F$ or $x_\alpha \notin F$. If $x_\alpha \notin F$, then from Bayesian monotonicity there exists $i, y : T^{-i} \to A$, and $t_i \in T^i$ such that

$$V^i(y_\alpha, t_i) > V^i(x_\alpha, t_i) \text{ and}$$

$$V^i(x, t_i') \geq V^i(y, t_i') \text{ for all } t_i' \in T^i.$$

For every (x, α) pair such that $x_\alpha \notin F$ define a 'test agent', $i(x, \alpha)$ and a 'test allocation rule' that is independent of i's type, $\hat{y}(x, \alpha)$: $T^{-i} \to A$ which satisfies the above inequalities. The mechanism is constructed as follows.

The message space of agent i is:

---

[18] This assumption is not needed, and is not made by Matsushima [1990b].

[19] A social choice correspondence satisfies *No Veto Power* if $a \in F(t)$ whenever $b^i(t^i) = a$ for at least $I - 1$ agents.

$$M^i = T^i \times \mathbb{A} \times F \times \{0, 1, 2, \ldots\} \times A$$

where $\mathbb{A}$ is the set of all possible joint deceptions. That is, each agent submits

(1) a reported type (m_1^i)
(2) a reported joint deception by the agents (m_2^i)
(3) an allocation rule in F (m_3^i)
(4) a non-negative integer (m_4^i)
(5) a feasible allocation (m_5^i)

We partition M into several regions and define the outcome function on these regions. Let $\bar{a} \in A$ be some fixed element of A, and let α_0 denote the identity joint deception. Let

$$D_1 = \{m \in M \mid \exists i, x \in F \text{ such that } (m_2^j, m_3^j, m_4^j, m_5^j) =$$
$$(\alpha_0, x, 0, \bar{a}) \forall j \neq i \text{ and } m_2^i = \alpha_0\}.$$

We call D_1 the *agreement region*. Let

$$D_2 = \{m \in M \mid \exists i, x \in F \text{ such that } (m_2^j, m_3^j, m_4^j, m_5^j) =$$
$$(\alpha_0, x, 0, \bar{a}) \forall j \neq i \text{ and } m_2^i \neq \alpha_0\}.$$

We call D_2 the *objection region*. Finally, the *disagreement region* is

$$D_3 = \{m \in M \mid m \notin D_1 \cup D_2\}.$$

In the agreement region, everyone reports 'no deceptions', and the last four components of at least $I - 1$ of the messages agree. In the objection region exactly one agent reports a deception and the remaining $I - 1$ messages are identical in their last four components. In the disagreement region there is no subset of $I - 1$ agents whose reports are identical in the last four components.

The outcome function is defined as follows:

$$g(m) = x(m_1) \qquad\qquad \text{if } m \in D_1$$

$$g(m) = \begin{cases} x(m_1) & \text{if } m \in D_2 \text{ and } i \neq i(x, m_2^i) \\ \hat{y}(x, m_2^i)(m_1^{-i}) & \text{if } m \in D_2 \text{ and } i = i(x, m_2^i) \end{cases}$$

$$g(m) = m_5^h \qquad\qquad \text{if } m \in D_3$$

where h is the agent submitting the highest integer (ties are broken in favor of the agent with the highest index). That is

$$H = \{j \in I \mid m_4^j \geq m_4^i \quad \forall i \in I\} \text{ and}$$

$$h = \max\{j \in H\}.$$

Thus, the outcome function ignores unilateral deviations from the agreement region (where the last four components of everyone's message are $(\alpha_0, x, 0, \bar{a})$, with one exception. That exception is where $i = i(x, \alpha)$ reports $(\alpha, ., ., .)$ and everyone else reports $(\alpha_0, x, 0, \bar{a})$ in the last four components. In that case the outcome changes from $x(m_1)$ to $\hat{y}(m_1^{-i})$. By construction, this 'selectively eliminates' the potential equilibrium outcome x_α. If everyone were always submitting $(\alpha_0, x, 0, \bar{a})$ in their last component, but reporting their first component deceptively, using deception α, then some type of $i(x, \alpha)$ could improve by unilaterally deviating to the objection region. The function $i(x, \alpha)$ was created to selectively eliminate *all* such potential equilibria.

All other joint strategies have the property that there is at least one agent, and at least one $t_i \in T^i$ where i deviates from $(\alpha_0, x, 0, \bar{a})$. The assumptions of BEPV and SCP always guarantee that at least one other agent can improve his expected payoff at some type $t_j \in T^j$ by reporting an integer higher than any integer reported by any type of any other agent,[20] and reporting $m_5^i = b^j(t_j)$. Therefore there can be no equilibria which involve less than complete agreement in the last four components of the message space. ■ ■ ■

B. Non-diffuse information structures

Jackson's [1991] class of economic environments is slightly more general than the one just presented, primarily because he allows for information structures which are not diffuse. There are several ways to represent the idea that information is not diffuse. In fact, the early work on Bayesian implementation by Postlewaite and Schmeidler [1986,1987] assumed an extreme form of nondiffuse information, which they called 'Non-Exclusive Information' (NEI). Roughly speaking, an information structure satisfies NEI if the type of agent i can be

---

[20] The finiteness of T is exploited here.

deduced with certainty from knowledge of the types of all agents other than i.

Postlewaite and Schmeidler formalized this idea without introducing the concept of a 'type'. In their formulation, there is an exhaustive set of *states of the world*, S, which we will take to be finite. For each i, we define a partition of the states, Π^i, and for each $s \in S$ denote $E^i(s)$ the unique element of Π^i containing s. Thus, $E^i(s)$ is the *event* i observes if the actual state is s, and it defines i's information set at s. Each individual has a prior, G^i defined over the set S, with the property that $G^i(s) > 0$ for all $s \in S$. If the true state is s, then i observes $E^i(s)$ and updates G^i according to Bayes' rule, so:

$$G^i(t|E^i(s)) = \frac{G^i(t)}{\displaystyle\sum_{\tau \in E^i(s)} G^i(\tau)} \quad \text{if } t \in E^i(s)$$

$$= 0 \qquad \text{if } t \notin E^i(s)$$

Individuals have state contingent utility functions, where $U^i(a, s)$ is the utility of allocation a to individual i in state s. The special case of complete information arises when $E^i(s) = \{s\}$ for all i and for all s. The case of NEI[21] arises when $\cap_{j \neq i} E^j(s) = \{s\}$ for all i and for all s. Thus, if all agents other than i could pool their information at s, they would know for sure that i must have observed $E^i(s)$. A social choice function is then just a mapping from S to A. A social choice correspondence (also sometimes called a *social choice set*) is just a collection of social choice functions. Postlewaite and Schmeidler identify a very basic condition that is always necessary for implementation of a SCC, F, called *closure*. Let Π denote the common knowledge partition implied by $(\Pi^1, \ldots, \Pi^I)$, and denote the common knowledge event at s by $E(s)$. Then F is *closed* if, for all x and x' in F, and for all E and E' in Π, $x'' \in F$, where x'' is defined by:

$$x''(s) = x(s) \text{ for } s \in E$$

$$= x'(s) \text{ for } s \in E'$$

It is easiest to see why this is a necessary condition for Bayesian

---

[21] Postlewaite and Schmeidler [1986] assume NEI. They also assume, essentially without loss of generality, that states are not redundant, in the sense that if the agents were to pool their information, they could identify the state exactly.

implementation by looking at the special case of complete information. Suppose there are only two states s and s', and the common
knowledge partition is $\{\{s\}, \{s'\}\}$ and F includes the following two
social choice functions:

$$f: \quad x(s) \quad = a \qquad f': \quad x'(s) \quad = b$$
$$ x(s') = b \qquad x'(s') = a$$

Closure of F requires that the following two social choice functions
must also lie in F:

$$g: \quad y(s) \quad = b \qquad g': \quad y'(s) \quad = a$$
$$ y(s') = b \qquad y'(s') = a$$

It is clear why closure is necessary for implementation. In this simple
example, suppose that F is implementable and that x and x' are both
in F, and let μ be a mechanism that implements F. Then it must be
that both a and b are Nash equilibria at s and both a and b are Nash
equilibria at s'. Thus the set of all social choice functions that choose
a or b in state s and choose a or b in state s' must also arise as
equilibrium outcomes of μ. Since F is implementable, they must lie in
F. This means that closure is satisfied. The argument for more general
information structures than complete information is the same.

Observe that, with complete information, closure implies that a SCC
can be represented *either* as a multivalued mapping (i.e. correspondence) from S to A *or* as a collection of functions from S to A. If
information is not complete, then there may be SCCs that satisfy
closure that cannot be represented as a multivalued mapping from S
to A. For this reason, with incomplete information it is necessary to
use the more general representation of a SCC as a collection of social
choice functions from S to A. Also observe that if information is
sufficiently diffuse then $\Pi = \{S\}$, in which case the closure requirement has no bite.

The state/partition/event representation of information structures
used by Postlewaite and Schmeidler is also used in Palfrey and
Srivastava [1987, 1989a]. This representation is equivalent to the 'type'
representation used in this monograph and in virtually all other papers
on Bayesian games and Bayesian mechanism design and Bayesian
implementation theory.

To illustrate the equivalence, consider first the case of complete information with two individuals, 1 and 2, each of whom could be one of two possible types. We can then define four possible states, one for each possible type profile. Thus we have

$$S = \{s_1 = (t_1^1, t_1^2), s_2 = (t_2^1, t_1^2), s_3 = (t_1^1, t_2^2), s_4 = (t_2^1, t_2^2)\}$$

where superscripts denote individuals. We then consider any partition such that each individual's partition is measurable in his own type, i.e. each individual knows his own type $(s_1 \neq s_2 \Rightarrow E^1(s_1) \neq E^1(s_2))$. Complete information requires that $E^i(s) = \{s\}$ for all i and s.

As a second illustration, consider diffuse information. This means that no individual has any information beyond his original G^i and the observation of his own type (i.e. $\Pi^1 = \{\{s_1, s_3\}, \{s_2, s_3\}\}$). Notice that this does not mean that types are statistically independent or that an individual's observation of his own type provides no information on the other agent's type. Suppose that $G^1 = (.1, .4, .4, .1)$. Then if 1 observes that he is type t_1^1, his posterior on agent 2 being type t_1^2 is .2, but if 1 observes that he is type t_2^1, his posterior on agent 2 being type t_1^2 is .8. What diffuseness does require is that all components of G^i are positive for all agents, that is all agents believe that any type profile can occur with positive probability, and that every event of every type of every agent includes all possible type profiles of the other agents.

Jackson [1991] represents nondiffuse information structures using the 'type' representation in the following way. Let T be the set of all type profiles, and let $\dot{S} \subseteq T$ be a subset of the type profiles with the property that, for all i, $G^i(s) > 0$ for all $s \in S$ and $G^i(t) = 0$ for all $t \notin S$.[22]

In the representation used by Postlewaite and Schmeidler, this means that the set of states, S, would be a subset of the set of all type profiles. Thus, the social choice functions defined in such a representation would not map T into A, but would map S into A. Jackson points out that it is much simpler to define social choice functions from T into A, and then work with equivalence classes of social choice functions which are defined by equivalence on their

---

[22] Notice that to keep matters relatively simple, we do not allow the possibility that the set of type profiles that some agents believe are impossible are different from the set of type profiles that other agents believe are impossible. Also notice that if $S = T$ then the information structure is diffuse.

restriction to S. Thus, given S, the social choice functions x and x' are *equivalent* if $x(s) = x'(s)$ for all $s \in S$. Two SCCs F and F' are equivalent if for all social choice functions x we have $x \in F$ if and only if there exists $x' \in F'$ with x' equivalent to x.

For economic environments satisfying SCP with at least 3 agents, Jackson gives the following result.

Theorem 3.5: Suppose $N > 2$ and SCP holds. The SCC F is implementable if and only if there exists an equivalent SCC, F', that satisfies closure, incentive compatibility; and Bayesian monotonicity.

Jackson [1991] also points out that this theorem immediately implies, as a corollary, the central finding of Palfrey and Srivastava [1989a]. In pure exchange economies with free disposal (in the sense that the 0 allocation is feasible), a social choice correspondence is implementable if and only if the equivalent correspondence that assigns the 0 allocation for all $t \notin S$ satisfies closure, incentive compatibility,[23] and Bayesian monotonicity.

'Non-economic' environments

The reason that economic environments are relatively easy to work with is that individuals disagree about most preferred elements. It has been known since the early work of Maskin [1977] that with complete information, in domains where fewer than $I - 1$ agents agree on a best element and $I > 2$, then implementation is fully characterized by monotonicity. The above findings for economic environments, summarized above, indicate that essentially the same kind of result holds for asymmetric information, with the additional requirement of incentive compatibility.

With complete information, a fairly general statement of this result is possible, by imposing the requirement of no veto power on the SCC. Recall that No Veto Power (NVP) says an allocation that is most preferred by at least $I - 1$ agents at state s must lie in $F(s)$. This condition is used in the sufficiency arguments to show that equilibrium outcomes that arise using nuisance messages (i.e. equilibria in the 'indirect' part of the message space) must satiate at least $I - 1$ agents.

---

[23] For the 'if' part of this corollary Palfrey and Srivastava [1989a] imposed a slight strengthening of incentive compatibility, which Jackson's theorem shows is unnecessary.

The situation is more complicated with asymmetric information for at least three reasons. First, the statement of NVP will be more complicated since F cannot generally be represented as a multivalued function from T to A. Second, with asymmetric information, an individual does not know for sure whether the other agents are using nuisance messages. Other agents may sometimes be using nuisance messages and sometimes not, depending on their types. Third, individuals may not have enough information to know their best element. Jackson [1991] shows that these additional complications mean that one cannot simply add a new condition similar to NVP to the three conditions (closure, incentive compatibility, and monotonicity) already used for economic environments. Instead, he proposes a new condition that blends together the key features of NVP and monotonicity. The condition is called *Monotonicity-No-Veto*.

Two preliminary definitions are needed. Given two social choice functions x and x', and some set of type profiles $C \subseteq T$, define a new social choice function, $xx'_{(C)}$, called the *splicing of x and x' along C*, by

$$xx'_{(C)}(s) \quad = x(s) \quad \text{if } s \in C$$
$$= x'(s) \quad \text{if } s \notin C$$

An allocation a satisfies the *No Veto Hypothesis* (NVH) at $t \in T$ if a is the most preferred allocation at t for at least $I - 1$ agents. Let $B^i \subseteq T^i$ and $B = B^1 \times \ldots \times B^I$. A social choice function satisfies *Monotonicity-No-Veto* if for every such B and every deception α and for every social choice function z not equivalent to x, such that $z(s) = x_\alpha(s)$ for $s \in B$ and $z(t)$ satisfies NVH at all $t \in S - B$, there exist i, $y: T^{-i} \to A$ and $t^i \in B^i$ such that:

$$y_\alpha z_{(B)} P^i(t^i) z \quad \text{and} \quad xR^i(\tau^i)y \quad \text{for all } \tau^i \in T^i.$$

Notice that this condition implies Bayesian monotonicity. To see this, set $B = T$. Jackson [1991] also shows that it reduces to the separate conditions of No Veto Power and Monotonicity when there is complete information. The definition for SCCs is defined accordingly, and leads to the following sufficiency result.

Theorem 3.6: If $N > 2$ and F satisfies closure, incentive compatibility and monotonicity-no-veto, then F is implementable.

A proof of this theorem appears in Jackson [1991]. The theorem represents the most general characterization of Bayesian implementable allocation rules that has been obtained so far. Recently, Dutta and Sen [1991] have shown that some of Jackson's results are easier to state (and can be extended) if one assumes diffuse information. In that paper, they identify a sufficient condition for implementation that is strictly weaker than Jackson's if information is diffuse. In addition, they obtain some results for two-person environments. This completes our discussion of the various characterizations of implementable SCCs, and we end this section with a brief discussion of some open issues in this area.

As discussed in the introduction, the Bayesian implementation problem is motivated by the desire to investigate whether socially desirable allocations can be achieved in a decentralized manner. The results we have described provide a precise characterization of when this is possible. In the next chapter we investigate some applications of these results.

The results summarized in this section are existence theorems, not applications. They identify conditions under which there will exist a mechanism whose Bayesian equilibria exactly correspond to the outcomes prescribed by a social choice function. It is not the case that the implementing mechanisms used in the proofs should be interpreted as being useful in specific applications. Quite to the contrary, they are very cumbersome. But the point of these proofs is simply to find some mechanism that works.

With these results in hand, if we find that some allocation rule violates the necessary conditions for Bayesian implementation, then we now have a clear understanding of what this means. It means that there is *no* institutional arrangement which will decentralize the SCC, if individuals follow Bayesian equilibrium behavior. However, if we obtain a positive result, i.e. we find that some appealing SCC satisfies the conditions described in this chapter, we do not yet have a theory of institutions. All we know is that there is *some* institutional structure which can be imposed and which will lead to desirable outcomes.

It cannot be emphasized enough that the constructions used in the sufficiency arguments are very abstract, and do not necessarily correspond to institutional arrangements we might see in practice. This is not surprising given that the theorems are very general and require the construction of a single mechanism to apply to virtually all environments.

One important open problem is to see which allocation rules can be implemented by 'realistic' mechanisms. One suspects (and hopes) that in many specific applications, relatively simple mechanisms, resembling ones we actually observe in practice, might work. In fact, the extensive literature on auction theory and bargaining mechanisms indicates that this is often the case.

A second, related problem is that all the results described are parametric in the sense that they require detailed knowledge of the set of possible types, the set of alternatives, and the prior beliefs of the agents. In practice, one might not expect such information to be readily available.

This raises two sub-issues. First, such information may not be readily available to the planner. In this case, it is very hard to justify parametric mechanisms.[24] Thus, another interesting open question is whether there exist non-parametric mechanisms which decentralize desirable SCCs. Second, to the extent that players themselves are not behaving as Bayesians with well-formed priors, the notion of Bayesian equilibrium itself must be brought into question. With the exception of the relatively negative results *vis-à-vis* dominant strategy equilibrium, there has been little exploration of 'non-Bayesian' approaches to implementation with incomplete information.

This seems to be an important direction to explore. We have examined the implementation problem assuming that all agents adopt Bayesian-Nash behavior, which requires a great deal of ability on the part of the agents. While we will discuss other concepts of equilibrium in Section 6, these will assume similar abilities. There has been no work to date on bounded rationality approaches to the implementation problem, and very little on the related issue of 'simple' mechanisms. This may well turn out to have close ties with the issues of realistic mechanisms and nonparametric mechanisms, discussed above.

Finally, we also know very little about the implementation problem when the set of feasible alternatives is privately known to some agents. An exception is the work of Hurwicz, Maskin and Postlewaite [1980] on complete information implementation in exchange economies when the planner does not know the endowments. In that analysis it is critical to prohibit agents from overstating their endowments. It is unclear how to extend their results to more general environments, such as more general alternative sets or diffuse information structures.

---

[24] For example, if the planner does not know the players' conditional priors about each others' types, it will usually not be possible to construct the 'test allocation rules', since they are typically prior dependent.

C. Bibliographic note: implementation with complete information

There is a substantial literature on the implementation problem when there is no incomplete information among the agents. In fact, a great deal of the early work on implementation theory took this approach. Most of this literature can be viewed as proceeding under the assumption that none of the players have any information that the other players do not have. That is, at the time the mechanism is played, there is no incomplete information between the players. However, the planner (or the court who will be overseeing the play of the mechanism) does not have any information about the players. For example, in a bargaining game between a litigating plaintiff and a (possibly liable) defendant, it may be the case that both the plaintiff and the litigant know exactly what the state of the world is (such as the degree of contributory negligence or the dollar value of the damages), but the court does not know. Nevertheless, the court wishes to enforce a legal solution which depends upon the value of the variables that it cannot observe directly. A problem such as this may be viewed as a problem of implementation with complete information.

In this bibliographic note we briefly describe how the Nash implementation approach, in which information is not explicitly modelled, ties in with the Bayesian approach as a special case with complete information. Surveys of this literature include Maskin [1985], Moore [1991], Postlewaite [1985], Groves [1982], Dasgupta, Hammond and Maskin [1979], Laffont and Maskin [1982].

The analysis of implementation with complete information is essentially the same as the analysis we have presented in this section, but there is one important simplification: the priors of the players are degenerate. Specifically, all players know the entire type profile. That is, $G^i(t_{-i}|t_i)$ is either 0 or 1 for every i, t_i, and t_{-i}. The most commonly analyzed problem in this approach then makes a further simplification that we have been making throughout this monograph, namely, that the set of feasible alternatives does not depend on the type profile. Thus, the problem reduces to associating each type profile with a preference profile, and then assuming that this preference profile is common knowledge among the players at the time they play a mechanism. This means, for example, that according to the definition given earlier in this chapter, a 'direct game' will involve all of the agents reporting an entire profile of preferences, from the domain of possible profiles.[25] Because type profiles are associated directly with preference profiles, it is common practice to write a social choice function as a mapping from the domain of profiles to A, rather than from a type space to A. If the profile is $R = (R_1, \ldots, R_I)$, then we write the social choice set at R as $F(R)$, where R_i is a complete and transitive ordering on A.

Several results become immediate. First, in many cases of interest, incentive compatibility is automatically satisfied. For example, if there are three agents, then for any social choice function, f, there exists a very simple direct mechanism that will weakly implement f. Pick an arbitrary alternative a^*. Every agent reports a type profile. If at least $(I - 1)$ of the I agents report R, then the outcome is $f(R)$. Otherwise, the outcome is a^*. Obviously it is an equilibrium at R for everyone to report R. The two-agent problem poses some special difficulties, and an incentive compatibility condition is also required. This stems from the fact that with three or more agents, it is possible to identify a unilateral deviation. For example, if two agents report R and the third reports R', then the third can be viewed as the deviator. With two agents, this is not possible. An extensive discussion of the restrictions imposed by incentive compatibility if there are 2 agents can be found in Dutta and Sen [1991], Moore and Repullo [1990], or Mookherjee and Reichelstein [1990b].

Thus, the first simplification in these complete information environments is that incentive compatibility is nearly always satisfied. Thus the only impediment to implementation is the multiple equilibrium problem. This is easily illustrated in the simple direct game in the previous paragraph. Not only is everyone announcing R an equilibrium when the true profile is R, but everyone announcing R is always an equilibrium for every R' in the domain.

Because complete information is formally just a special case of a Bayesian environment like the ones described earlier in this chapter, it follows that a necessary condition for Nash implementation is a simple version of Bayesian monotonicity. It is easy to show that in the special case of complete information, the relatively complicated condition given in Definition 3.1 can be written simply as follows.

---

[25] This contrasts with the original use of the term 'direct' mechanism to refer to a report of one's private characteristics (rather than a report of everything that one knows). Part of the apparent confusion here is that the original application of this term was used in the context of dominant strategy mechanisms, where there was no assumption of complete information. See, for example, Green and Laffont [1977].

Definition 3.1 (monotonicity): A social choice correspondence F is *monotonic* if, for all R, R', if:

1) $x \in f(R)$

and

2) for all $y \in A$, $i \in I$:
$$xR_i y \Rightarrow xR_i'y$$

then:

$$x \in f(R')$$

One can then prove results analogous to Theorems 3.1(i) and 3.5:

Theorem 3.1': If a social choice correspondence is Nash implementable, then it is monotonic.

Theorem 3.5': With complete information, if $I \geq 3$ and SCP is satisfied, then a social choice correspondence, F, is Nash implementable if and only if F is monotonic.

This is a simple corollary of a result that was first stated by Maskin in his seminal 1977 paper, and for which a complete proof has since been given by Williams [1984], Saijo [1988], Repullo [1987], and others. The assumption of strongly conflicting preferences serves the same role as No Veto Power (see footnote 19). This gives us:

Theorem 3.7: With complete information, if $I \geq 3$, and a social choice correspondence, F, is monotonic and satisfies No Veto Power, then F is Nash implementable.

A very simple constructive proof is adapted from Repullo [1987]. A typical element of the message space consists of an element of A, a reported preference profile and a nonnegative integer. If the joint message is such that there exists some i some R and some $a \in F(R)$, such that everyone except i sends the message $(a, R, 0)$, then the outcome is either a or a_i, where a_i denotes the element of A in player i's message: it is a_i if and only if $a R_i a_i$. For any other joint message, the outcome is a_{j*}, where j^* is the player who reported the highest integer (ties are broken in favor of the player with the lowest index).

This basic result has been extended recently by Moore and Repullo [1990], Danilov [1992], Yamato [1990a], and by Dutta and Sen [1991]. The first of these provides a necessary and sufficient condition for a social choice correspondence to be Nash implementable, and the last of these provides a similar characterization for the 2-person case.

Danilov [1992] gives a very simple necessary and sufficient condition for Nash implementation for the case of 3 or more agents. Yamato [1990a] provides a slight extension and simplification of Danilov [1992].

There are also some characterization theorems for specific environments, such as pure exchange economies. Much of that literature is summarized in the Postlewaite [1985] survey. Hurwicz, Maskin and Postlewaite [1980] provide an extensive analysis of the problem of implementation with complete information in exchange economies when the planner does not know initial endowments (so feasibility is an issue).

With complete information, there has been a lot of attention focussed on implementation with refinements of Nash equilibrium. This makes sense, since the main impediment to implementation with complete information is the multiple equilibrium problem. While it has been known for some time that sequential rationality (for example, subgame perfection or dominance solvability) expands the set of implementable allocations with complete information,[26] it was not until the influential paper of Moore and Repullo [1988] on subgame perfect implementation that the enormous power of refinements was revealed. There they show that in many environments with some form of infinitely divisible private good all social choice functions satisfy a condition that they prove is sufficient for implementation in subgame perfect Nash equilibrium. They provide a constructive proof using stage games. Their result was generalized by Abreu and Sen [1990], where a more general characterization for the case of 3 or more agents is proved.

If there is no divisible private good, then there are many social choice correspondences that cannot be implemented using the stage game

---

[26] This is evident, for example, in the specific applications examined by Moulin [1979] and Crawford [1980], as well as the earlier work on sophisticated voting by Farquharson [1957/69], even though this last reference did not focus primarily on implementation issues.

approach, because subgame perfection fails to eliminate equilibria that are implausible when players use weakly dominated strategies. For example, virtually all voting rules, such as plurality voting, the Borda count, or the Hare runoff system, fail to satisfy the necessary conditions for subgame perfect implementation in environments without transfers.

Palfrey and Srivastava [1991a] introduce an alternative refinement for normal form games, called *Undominated Nash Equilibrium*. This refines away all Nash equilibria in which some player uses a weakly dominated strategy. (It does not iterate this elimination of weakly dominated strategies.) They find that essentially every social choice correspondence satisfying No Veto Power is fully implementable in undominated Nash equilibrium. A similar result holds for implementation in trembling-hand perfect equilibrium with finite alternatives and strict preferences (Sjostrom [1990c]). Jackson [1989] proves that a similar construction to Palfrey and Srivastava [1991a] works for the very weak solution concept of *Undominated Equilibrium* (where σ is an equilibrium if and only if σ_i is not weakly dominated for each i).

The Jackson [1989] paper raises an important issue. In the constructive proofs used by Palfrey and Srivastava [1991a], the mechanism used to prove the theorem has an unattractive feature. In particular, some equilibria are eliminated by giving a player an infinite collection of messages, with the property that they form an infinite chain of weakly dominated strategies. This has been called the 'tailchasing' technique: one of the messages in this set is used to prevent some other message outside the set from being an equilibrium, and then the undominated refinement prevents this new message from being part of an equilibrium message profile. Both Palfrey and Srivastava [1991a] and Jackson [1989] offer examples which suggest that the undominated refinement may be inappropriate to apply to games with this feature.

The reason this paper by Jackson is important is that it is the first paper that suggests placing axiomatic restrictions on the class of mechanisms that can be used.[27] His restriction is that if some strategy

---

[27] It should be pointed out that in many cases — including this one — restrictions on the mechanism can be restated as an equilibrium concept. Nevertheless, it makes more intuitive sense to think of this approach at placing limitations on games, rather than placing limitations on equilibria.

is weakly dominated, it must be weakly dominated by some strategy that is not itself weakly dominated by some other strategy. He calls such mechanisms *bounded*. He shows that, the only social choice functions that are implementable in undominated equilibrium by bounded mechanisms are strategyproof. Thus, at least for his equilibrium concept (which is *not* a refinement of Nash equilibrium), we go from one extreme of implementing everything with unbounded mechanisms to the opposite extreme of implementing almost nothing with bounded mechanisms.

This discussion leads to the following question: what can be implemented in undominated Nash equilibrium using bounded mechanisms? A partial answer was given in Palfrey and Srivastava [1991a], who showed that the combination of strict preferences, No Veto Power, and a unanimously worst outcome allowed a bounded construction. Jackson, Palfrey, and Srivastava [1990] extend this and identify some weak necessary conditions and some sufficient conditions for bounded implementation in domains with 3 or more agents when no veto power is satisfied. They show that most social choice correspondences of interest can be implemented using bounded mechanisms. They also provide a very simple construction to implement social choice functions in exchange economies using a dominance solvable mechanism.

Dominance solvable implementation has been investigated in the context of stage games, by Herrero and Srivastava [1992], and in normal form games by Moulin [1979] and by Abreu and Matsushima [1990a]. The latter paper uses a construction developed from their earlier papers on virtual implementation.

A social choice function is virtually implementable as long as some ε-close social choice function is implementable. The first paper written on virtual implementation is Matsushima [1988]. A more accessible paper which significantly extends his results is Abreu and Sen [1991]. A pair of closely related papers by Abreu and Matsushima [1990a, 1992] has investigated virtual implementation by iterative removal of strictly dominated strategies. They find that nearly everything can be implemented in this way as long as there are at most a finite number of (von Neumann–Morgenstern) preference profiles in the domain, and there are conflicting preferences (for example a transferable private good).

The above summary leaves out a good deal. This would include: applications by Glazer and Ma [1989] and Austen-Smith and Banks

[1991]; the pioneering work on public goods mechanisms by Hurwicz [1979, 1980] and Groves and Ledyard [1977, 1979]; recent work on renegotiation-proof implementation by Green and Laffont [1987b], Aghion, Dewatripont, and Rey [1989], Maskin and Moore [1989], Rubinstein and Wolinsky [1991]; a paper on 'double implementation' by Yamato [1990]; and more. In fact, the implementation problem with complete information has spawned an even more extensive literature than Bayesian implementation, no doubt because the absence of incentive compatibility complications greatly simplifies the analysis of the multiple equilibrium problem.

4. APPLICATIONS

In this section, we investigate the extent to which the multiple equilibrium problem discussed in the previous chapter arises in the types of models studied by economists. As we will see, the examples in the previous chapter are not only highly illustrative of basic concepts, but actually quite representative of the kinds of problems that arise in these environments. In addition, we investigate the extent to which the multiplicity problem can be resolved by using indirect mechanisms.

A. Pure exchange economies

We start by examining whether some interesting social choice correspondences are implementable in pure exchange economies. Suppose there are L commodities, and let $w^i \in \mathbb{R}^L_+$ denote the endowment of agent i. Let $w = \Sigma_i w^i$ denote the aggregate endowment. Suppose that w^i is independent of the type of agent i, so that w is also independent of the types of the agents. Then, $A = \{a \in \mathbb{R}^L_+ \,|\, \Sigma_i a^i \leq w\}$ is the feasible set of alternatives, and given $a \in A$, $U^i(a^i, t)$ denotes the *ex-post* utility of agent i at type profile t. As implicit in the notation, agent i is only concerned with his own allocation of commodities. We further assume that for each t, U^i is strictly increasing and strictly concave in the first argument.

1. Example 4.1

Consider a pure exchange economy with two goods, two agents, and an aggregate endowment w. An allocation is a pair of consumption bundles, $x_1 = (x_1^1, x_1^2)$ and $x_2 = (x_2^1, x_2^2)$ where x_i^j denotes agent i's allocation of good j. Feasibility simply requires that $x_1 + x_2 \le w$. Agent 1 has preferences represented by utility function u_1 or u_1' with probability p and $(1 - p)$, respectively. Agent 2's preferences are commonly known to be u_2. Thus there are only 2 possible vectors of types, say t and t'; agent 1 is fully informed; agent 2 is uninformed but knows p. Figure 4.1 illustrates a particular *ex-post* efficient allocation rule $(x(t), x(t'))$. However this allocation is not the only equilibrium in the corresponding direct mechanism. In particular, the informed agent is indifferent between announcing 't' and announcing 't'' when the true type is t'. But this equilibrium produces the allocation, $(x(t), x(t))$, which is not *ex-post* efficient (in particular, $(x(t), x(t'))$ dominates $(x(t), x(t))$.

However, there is a simple indirect mechanism which implements x. Consider adding a strategy, call it 'N', for agent 2, the uninformed agent, and modify outcomes as follows. If agent 2 does not say N, then the outcome is $x(t)$ if agent 1 says t, and $x(t')$ if agent 1 says

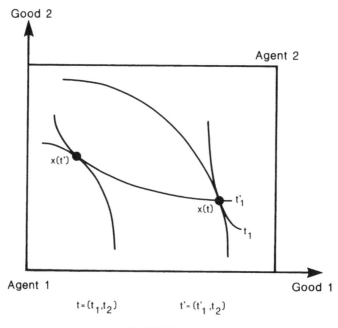

FIGURE 4.1

t'. If agent 2 says N, then the outcome is y if agent 1 says t and $x(t)$ if agent 1 says t', where y is a point in Figure 4.1 on player 1's t'-indifference curve such that:

$$U_2(y) = \text{Min}\{U_2(x(t')), U_2(x(t)) + \frac{1-p}{p}[U_2(x(t')) - U_2(x(t))]\}$$

Now, it is not an equilibrium for agent 1 to always say t; if he does, agent 2 should announce N, and agent 1 is now worse off at t. It can be verified that all equilibrium outcomes to this modified mechanism produce x.[28] It can also be checked that Bayesian monotonicity is satisfied. In particular, if $\alpha^1(t) = \alpha^1(t') = t$, then $y(t) = y, y(t') = x(t)$ will satisfy the required inequalities for this α.

2. Market equilibria
An important result in the theory of implementation with *complete* information relates to the Walrasian correspondence. It is known that except for boundary problems, the Walrasian correspondence is implementable when agents have complete information.[29]

With incomplete information, one concept of market equilibrium is that of a Rational Expectations Equilibrium (REE). A natural question to ask is whether the allocation rule produced by an REE is implementable. This question has been analyzed by Blume and Easley [1983, 1990], Palfrey and Srivastava [1987], Wettstein [1986], Chakravorti [1992], and others. An REE allocation rule is defined as follows.

Let $P(t)$ denote the vector of prices when the vector of types is t. Given a price function $P: T \rightarrow \mathbb{R}_+^L$, let

$$E^i(t_i, p; P) = \{(t'_{-i}, t_i) \in T \mid P(t'_{-i}, t_i) = p\}$$

denote the set of type profiles which i cannot distinguish based either on his own type or the price. Let $G^i(t_{-i} \mid t_i, p; P)$ be the distribution function of agent i, conditional on $t \in E^i(t_i, p; P)$. This distribution

---

[28] Notice that if $p < .5$, then $y = x(t')$, in which case there are two equilibria: agent 1 reports truthfully and agent 2 does not say N; and agent 1 always lies while agent 2 reports N. If $p > .5$, then the solution is still fairly easy. In fact the second equilibria vanishes, since 2 will no longer wish to report N when agent 1 always lies.

[29] This result requires that there be at least three agents. The boundary problems also arise with incomplete information, and are discussed below.

function incorporates both the private information of agent i and the public information contained in the price vector. For example, if P is fully revealing, so $P(t) \neq P(t')$ if $t \neq t'$, then $E^i(t_i, p; P)$ is a singleton (or empty) for each (t_i, p) pair, and $G^i(t_{-i}|t_i, P(t); P) = 1$ when t is the true vector of types and 0 elsewhere.

Given a price function P, let $\xi^i(t_i, p; P)$ denote the demand function of type t_i of agent i faced with prices p. If $E^i(t_i, p; P) \neq \phi$, then $\xi^i(t_i, p; P)$ maximizes $\int U^i(a^i, t) dG^i(t_{-i}|t_i, p; P)$ subject to $p \cdot [a^i - w^i] \leq 0.^{30}$ Note that the support of G^i equals $E^i(t_i, p)$, so prices are equal across all $t \in E^i(t_i, p)$. If $E^i(t_i, p; P) = \phi$, then define $\xi^i(t_i, p; P) = w^i$.

A *Rational Expectations Equilibrium* is defined to be a price function, P, such that for each t, $\Sigma_i \xi^i(t_i, P(t); P) = w$. If P is an REE, the implied allocation rule, x, is defined by $x^i(t) = \xi^i(t_i, P(t); P)$. Let $F_R : T \to A$ be the REE correspondence, i.e. $F_R = \{x : T \to A \mid \exists P$ such that P is an REE and x is the implied allocation rule$\}$. The question is whether F_R is implementable.

As shown in the previous chapter, in order for F_R to be implementable, every $x \in F_R$ must be incentive compatible. In addition, Bayesian monotonicity must be satisfied. Blume and Easley [1990] show that REE correspondence usually does not satisfy incentive compatibility. The following example, taken from Palfrey and Srivastava [1986], illustrates the problem.

Example 4.2: There are two agents $T^1 = \{t', t''\}$ and agent 2 has only one type. There are two commodities with $w^1 = (2, 0)$, $w^2 = (1, 1)$. Preferences are as follows:

$$U^i(a, t) = \beta^i(t) \log(a_1^i) + [1 - \beta^i(t)] \log(a_2^i)$$

where

$$\beta^1(t') = \beta^1(t'') = 2/3$$

$$\beta^2(t') = 1/4, \beta^2(t'') = 3/4.$$

Then, the following is an REE price function:

$$P_1(t') = 3/17, p_2(t') = 1, p_1(t'') = 9/11, p_2(t'') = 1,$$

---

[30] The assumptions of strict concavity guaranteed that there is a unique maximum.

and leads to the following allocations for agent 1:

$$x_1^1(t') = 4/3, x_2^1(t') = 2/17, x_1^1(t'') = 4/3, x_2^1(t'') = 6/11.$$

It can be verified that agent 1 is always better off reporting type t'' no matter what his true type.

Thus, we see that F_R does not satisfy incentive compatibility. Blume and Easley [1990] present an example in which every $x \in F_R$ violates incentive compatibility. In the example above, there is also a non-revealing equilibrium (i.e. one with $p(t') = p(t'')$) which is incentive compatible.

It turns out that the problem with implementing the REE correspondence stems primarily from the fact that REE allocation rules do not satisfy incentive compatibility. Palfrey and Srivastava [1987], study the implementability of F_R with non-exclusive information, so that incentive constraints are not binding. They show that if utility functions are bounded away from the boundary of A, and endowments are state independent, then F_R is implementable.

The assumption on utility functions ensures that all REE allocations are strictly interior to A, and relaxing this assumption leads to exactly the same problem as in the complete information case (see Hurwicz, Maskin, and Postlewaite [1980]).

B. Efficient allocations

1. Pure exchange economies

In the Example 4.1, we saw that an efficient allocation rule could not be implemented by a direct mechanism, but could be implemented by an indirect mechanism. With complete information, it is known that in pure exchange environments, the social choice correspondence defined by the set of all Pareto optimal allocations is implementable (in Nash equilibrium). An important question in pure exchange economies is to see if this result extends to economies with incomplete information in a natural way. This problem has been studied by Palfrey and Srivastava [1987]. Their results are negative; they show that generally, the analogous set of efficient allocation rules cannot be implemented. The following example, taken from Palfrey and Srivastava [1989a] demonstrates this for *ex-ante* efficient allocation rules.

Example 4.3:   There are two agents and 1 good. There are two possible types for each agent 1, $T^i = \{t, t'\}$, with $G(t) = 1/2 = G(t')$. The types are perfectly correlated, so either both agents are of type t or both are of type t'. Thus, at the interim stage, there is complete information. Preferences are as follows:

$$U^1(a^1, t) = \log(a^1), \; U^1(a^1, t') = 2\log(a^1),$$

$$U^2(a^2, t) = U^2(a^2, t') = \log(a^2).$$

Suppose $w = 5$. Then, the following allocation rule is *ex-ante* efficient:

$$x^1(t) = 5/2, x^1(t') = 10/3, x^2(t) = 5/2, x^2(t') = 5/3.$$

Consider the pooling deceptions $\alpha^i(t) = \alpha^i(t') = t$, so $x_\alpha(t) = x_\alpha(t') = x(t)$. It is easy to check that x *ex-ante* Pareto dominates x_α. However, it can be verified that there is no $y: T \to A$ which satisfies the requirements of Bayesian monotonicity for this α.

It can also be shown that neither the set of interim efficient allocation rules nor the set of *ex-post* efficient allocation rules can be implemented. Examples illustrating these observations usually rely on dependent types and on dependent values (as does the above example). If there is no divisible private good, examples can be constructed to show that it is impossible to implement efficient allocations even with private values and independent types. However, in some plausible domains (such as those with private values, independent types, and transferable utility), any efficient allocation rule is implementable. It is shown below that in the private-values independent types model efficient allocations are implementable if there is also a transferable private good.

2. General environments

We have seen that efficiency and implementation are incompatible in pure exchange economies. The next example shows that in more general environments, this result even holds in the private-values, independent types model. The example exhibits an allocation rule which is *ex-ante* efficient and is not implementable.

Example 4.3:   There are three agents, and $A = \{a, b\}$. Each agent has two possible types $T^i = \{t_a, t_b\}$ for all i. Preferences are given by:

$$U^i(a, t_a) = 1 > U^i(b, t_a) = 0$$

$$U^i(b, t_b) = 1 > U^i(a, t_b) = 0.$$

Thus, an agent of type t_a strictly prefers a to b, while type t_b strictly prefers b to a. Suppose that types are independently drawn with $q^i(t_b) = q$ for all i and $q^2 > .5$.

The allocation rule x defined in Figure 4.2 is *ex-ante*, interim, and *ex-post* efficient. Note also that at each t, $x(t)$ is the (unique) majority winner at t.

To see that x is not implementable, consider the following deception: $\alpha^i(t_i) = t_b$ for all i, so $x(\alpha(t)) = b$ for all t. Note that x_α is *not* efficient (x dominates x_α for any of the three definitions of efficiency).

To see that Bayesian monotonicity is not satisfied, note first that y_α is a constant allocation rule. Thus, we cannot have $y_\alpha(t) = b$, since then $x_\alpha = y_\alpha$, and the strict inequality required by Bayesian monotonicity could not be satisfied. Thus, we must have $y_\alpha(t) = a$. If there exists an agent i and a type for i, say t_i, such that the strict inequality is satisfied, we must have $t_i = t_a$ since a is the worst element for type t_b. Picking $y_\alpha(t) = a$ implies that the strict inequality is satisfied for every agent at type t_a. Suppose, without loss of generality, that $i = 1$. The expected utility from x at t_a is $1 - q^2$ while that from $y(t_{-1}, \alpha^1(t_a))$ at t_a is:

$$(1 - q)^2 \, U^1(y(t_b, t_a, t_a), t_a) + 2q(1 - q) \, U^1(y(t_b, t_b, t_a), t_a) + q^2 U^1(y(t_b, t_b, t_b), t_a).$$

Since $y(t_b, t_b, t_b) = a$, this reduces to

$$q^2 + (1 - q)^2 \, U^1(y(t_b, t_a, t_a), t_a) + 2q(1 - q) \, U^1(y(t_b, t_b, t_a), t_a)$$

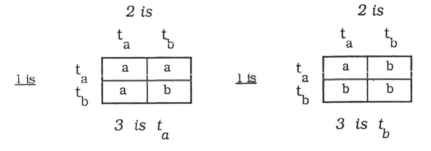

FIGURE 4.2

The minimum value of this last expression over y is q^2, which is greater than $1 - q^2$, so the inequalities required by Bayesian monotonicity must be violated when agent 1 is of type t_a. Hence, x is not implementable.

The fact that Bayesian monotonicity is not satisfied implies that there is no mechanism in the above setting which produces x as an equilibrium outcome, and which has no inefficient equilibrium outcome. This result, on the non-implementability of efficient allocations, stands in stark contrast to positive results which have been obtained with complete information.

3. Transferable utility models

We start with the private-values, independent types model with transferable utility. Let A be a finite[31] set of alternatives, $A = \{a_1, \ldots, a_m\}$, and denote by $\mathbb{P}(A)$ the set of all probability measures on A. If agent i is of type t_i, we write his *ex-post* utility as $U^i(a, t_i) + y^i$, where y^i is the transfer to i.

In this model, an allocation rule consists of a pair of functions, $p : T \to \mathbb{P}(A)$ and $y : T \to \mathbb{R}^1$. For each vector of types, the first function gives the probability of a particular alternative being chosen while the second specifies transfers between agents. The argument below relies on the concept of a *reduced form* allocation rule. Given functions p and y let

$$P^i(t_i) = \int dp(a, t_i, t_{-i}) dG(t_{-i}), \text{ and}$$

$$Y^i(t_i) = \int y^i(t_i, t_{-i}) dG(t_{-i}).$$

Then, $P^i(t_i)$ is the marginal distribution on A as a function of agent i's type, while $Y^i(t_i)$ is the expected transfer to i when of type t_i. The collection of functions $\{P^i(t_i), Y^i(t_i)\}$ are called *reduced form* allocation rules. The importance of these allocation rules, which are derived from the original allocation rule, stems from the following observation. Suppose that instead of being asked to report a type in the direct mechanism (p, y), agent i was told instead that the alloca-

---

[31] The analysis presented by Palfrey and Srivastava [1991b] does not require A to be finite, and also allows for more general forms of preferences.

tions would be chosen according to his set of reduced form allocations. Thus, if i reported the type t_i', an alternative would be chosen according to the distribution $P^i(t_i')$ and i would receive the transfer $Y^i(t_i')$. Then, it turns out that if (p, y) is incentive compatible, the agent i would report his type truthfully when asked to choose among his reduced form allocation rules. To see this, consider the expected utility to i when i is of type t_i and reports t_i':

$$v^i(t_i, t_i') = \int \left[\int U^i(a, t_i) dp(t_i', t_{-i}) + y^i(t_i', t_{-i}) \right] dG(t_{-i})$$

$$= \int \left[U^i(a, t_i) \int dp(t_i', t_{-i}) \right] dG(t_{-i}) + Y^i(t_i')$$

$$= \int U^i(a, t_i) dP^i(t_i') + Y^i(t_i').$$

Thus, the expected payoff to i if he reports t_i' is exactly the same as that which he would receive if he picked the reduced form allocation rule corresponding to t_i'. Since (p, y) is incentive compatible, it follows that if given a choice between the reduced form allocation rules, i would report his type truthfully.

With this observation, it is straightforward to establish the following result. Let $V^i(t_i)$ be the interim expected utility to type t_i of agent i under p, y, and given a mechanism (M, g), let $W^i(t_i, \sigma)$ be the interim expected utility to type t_i of agent i from an equilibrium σ.

Theorem 4.1: If (p, y) is interim efficient, then there exists a mechanism (M, g) such that if σ is an equilibrium to (M, g), then $W^i(t_i, \sigma) = V^i(t_i)$.

This result says that the interim *utility* allocations corresponding to an interim efficient allocation rule can be implemented. Palfrey and Srivastava [1991b] call this type of implementation *essential implementation*.

The proof of Theorem 4.1 is straightforward. Let $M^i = T^i \times [0, 1]$ for all i. If all agents report $(t_i, 0)$, then the outcome is $(p(t), y(t))$. If some agents report a positive number in the second component of their message space, then pick the agent reporting the lowest number, say i, and consider i's report, say (t_i, ε). Choose the alternative according to the distribution $P^i(t_i)$, and give i the transfer

$Y^i(t_i) - \varepsilon$. We can give the other agents any transfer we like. For example, we could 'balance' the transfers by dividing $Y^i(t_i) - \varepsilon$ equally among the other agents.

Note that since the reduced form allocation rules are incentive compatible, one equilibrium to this mechanism is for every agent to report his type truthfully and to report zero. Second, consider any equilibrium which yields type t_i of agent i expected utility strictly less than $V^i(t_i)$. Then, i can deviate to (t_i, ε) for some small enough ε, and get a payoff arbitrarily close to $V^i(t_i)$. Thus, if σ is an equilibrium, it must be the case that $W^i(t_i, \sigma) \geq V^i(t_i)$ for all i and t_i. Since (p, y) is interim efficient, we get that $W^i(t_i, \sigma) = V^i(t_i)$ for every equilibrium σ. ∎ ∎ ∎

This theorem also holds for mixed strategy equilibria. Since any agent can get arbitrarily close to $V^i(t_i)$, it follows that any mixed strategy equilibrium σ must also satisfy $W^i(\sigma, t_i) = V^i(t_i)$. Note that the proof exploits a discontinuity in the outcome function.

While we have shown that any interim efficient allocation rule is essentially implementable by constructing an implementing mechanism, it is also possible to prove the result by an application of the results summarized in Chapter 3, relating implementability to Bayesian monotonicity. Let

$$F = \{ (p_\alpha, y_\alpha) \mid V^i(p_\alpha, y_\alpha, t_i) = V^i(p, y, t_i) \text{ for some deception } \alpha,$$

$$\text{for all } i \text{ and } t_i \},$$

where $V^i(p, y, t_i)$ is the interim utility to i at type t_i from the allocation rule (p, y), and $V^i(p_\alpha, y_\alpha, t_i)$ is the interim utility to i if all agents employ the deception α. Thus, F is the set of deceptive allocation rules derived from (p, y) which are interim equivalent to (p, y). Then, (p, y) is essentially implementable if F is implementable. It is straightforward to see that if α is such that $(p_\alpha, y_\alpha) \notin F$, then there is an allocation rule, say (q, z), which satisfies the requirements of Bayesian monotonicity. If $(p_\alpha, y_\alpha) \notin F$, then there exists i and t_i such that $V^i(p_\alpha, y_\alpha, t_i) < V^i(p, y, t_i)$. Let $q(t) = P^i(t_i), z^i(t) = Y^i(t_i) - \varepsilon$, and define $z^j(t)$ in any manner. Then, for sufficiently small ε, (q, z) will serve to eliminate the deception α. A formal proof is given in Palfrey and Srivastava [1991b]. This argument is similar to that in Mookherjee and Reichelstein [1990a].

C. Bilateral monopoly

An extensively analyzed family of mechanisms for allocating an object between a single buyer and a single seller is the class of k-double auctions. According to a k-double auction, the buyer submits a bid, b, and the seller submits an offer, s. If $s > b$, then no sale occurs. If $s \leq b$, then the good is transferred from the seller to the buyer at a price of $p(s, b) = kb + (1 - k)s$.

It is by now well-known that the k-double auction has a plethora of equilibria — in fact, a continuum of equilibria. This multiplicity problem has drawn wide attention in the literature (see Leininger, Linhart, and Radner [1989], Matthews and Postlewaite [1989], Radner and Schotter [1989], Satterthwaite and Williams [1989], Williams [1987], Palfrey and Srivastava [1991b], and elsewhere.)

This problem is particularly troublesome since some of these equilibria are clearly undesirable. This is starkly illustrated in the special case of $k = 1/2$, and when valuations of the buyer (v_b) and seller (v_s) are independently distributed according to the uniform distribution on $[0, 1]$. Chatterjee and Samuelson derived one 'natural' Bayesian equilibrium for this case. Both agents adopt linear strategies given by

$$B(v_b) = \tfrac{1}{12} + \tfrac{2}{3} v_b$$

$$S(v_s) = \tfrac{1}{4} + \tfrac{2}{3} v_s.$$

Trade takes place if $v_b \geq v_s + 1/4$ and the buyer ends up paying $1/6 + 1/3(v_b + v_s)$. Remarkably, this allocation rule is *ex-ante* efficient (Myerson and Satterthwaite [1983]): it generates more expected surplus than any other feasible, incentive compatible, individually rational mechanism. This would seem to argue strongly for the use of such mechanisms in practice (at least when distributions of valuations are independent and approximately uniform).

Leininger, Linhart, and Radner [1989] show that there are *single-price equilibria*, where there is *some* value between 0 and 1 such that trade takes place (at price Z) if and only if $v_s \leq Z \leq v_b$. Furthermore, there are variations on these equilibria that involve step functions of prices at which trade may take place. Moreover, such equilibria are observed empirically (Radner and Schotter [1989]). As if this were not bad enough, Satterthwaite and Williams [1989] and Leininger,

Linhart, and Radner [1989] demonstrate that there is also a continuum of differentiable equilibria.

These multiplicity problems indicate an obvious role for indirect mechanisms – to eliminate the undesirable equilibria. Since this model is covered by Theorem 4.1 above, one way to implement the Chatterjee–Samuelson solution is to examine the direct mechanism corresponding to their equlibrium and apply Theorem 4.1. However, it turns out that it is possible to modify the double auction itself and eliminate all undesirable equilibria. The construction is similar to that used in Theorem 4.1. Consider letting each agent announce bids/offers as above and to also announce a number in [0, 1]. If no agent announces a positive number, then we follow the rules of the double auction. If one or more announce positive numbers, we pick the agent announcing the smallest strictly positive number (breaking ties randomly), and give him the reduced form allocation rule corresponding to the Chatterjee–Samuelson equilibrium, but charge him the positive amount he reported. Then, as in the proof of Theorem 4.1, it can be seen that all equilibria to this modified double auction yield exactly the same interim payoffs as in the Chatterjee–Samuelson equilibrium.

D. Incentive contracting with multiple agents

The next example is a modification of the model studied by Demski and Sappington [1984] and subsequently analyzed by Ma, Moore, and Turnbull [1988].

The model is that of a principal who owns two production technologies. There are two agents, and each agent can operate at most one technology. Operation of the technology requires effort on the part of agents, and it is assumed that this effort cannot be monitored. There are two possible types for each agent, $T^i = \{1, 2\}$ for both i, and we denote by a^i the action taken by an agent. If an agent expends effort a^i, receives compensation y^i and is of type t_i, the *ex-post* utility of the agent is $U^i(a^i, y^i, t_i) = y^i - t_i a^i$. Suppose that the types of the agents are distributed as follows. The probability of both being type 1 is 0.4, that of both being the second type is 0.4, while each probability that they are of different types is 0.1.

The principal needs to choose a compensation scheme for each agent and a recommended action level for each agent. Let y_{ij} be the payment to the first agent when the first agent is type i and the second agent is type j, and let a_{ij} denote the recommended action in the same setting. Then, incentive compatibility[32] for agent 1 when of type 1 requires

$$0.8(y_{11} - a_{11}) + 0.2(y_{12} - a_{12}) \geq 0.8(y_{21} - a_{21}) + 0.2(y_{22} - a_{22}).$$

When this agent is of type 2, incentive compatibility requires

$$0.8(y_{22} - a_{22}) + 0.2(y_{21} - a_{21}) \geq 0.8(y_{12} - a_{12}) + 0.2(y_{11} - a_{11}).$$

If we normalize the utility of each agent if unemployed to zero, then the participation constraint requires the left-hand side of each of these constraints to be non-negative.

The assumptions on technology are that there are decreasing returns to effort, and that a type 1 agent is strictly more productive than a type 2 agent. Under these assumptions, Ma, Moore, and Turnbull [1988] show that if a set of payments and action recommendations maximize the principal's *ex-ante* payoff subject to incentive compatibility and participation, then the second incentive constraint holds with equality, the type 2 agent receives a zero payoff, and[33] $y_{11} \geq y_{12}$. If we further assume that the production technology and the payoffs to the principal are such that $a_{11} = a_{12} = 2$ and $a_{21} = a_{22} = 1$ at the optimum, we get $y_{22} = y_{21} = 2$ and $0.8y_{12} + 0.2y_{11} = 4$.

The problem of minimizing the cost to the principal subject to the above constraints and given the optimal action choices then reduces to:

Minimize $0.4y_{11} + 0.1y_{12}$ subject to $y_{11} \geq y_{12}$ and $0.8y_{12} + 0.2y_{11} = 4$.

The solution is to set $y_{11} = y_{12} = 4$, which means that the interim payoff to a type 1 agent is 2, while that to a type 2 agent is zero.

The optimal contract, or direct mechanism, is summarized in Figure 4.3. The first row of each entry denotes the effort level of each agent while the second row denotes the compensation. Thus, if agent 1 reports type 2 while agent 2 reports type 1, then agent 1 works at level 1 and is paid 2, while agent 2 works at level 2 and is paid 4.

The problem with this contract is that the type 2 agent is indifferent between being truthful and pretending to be a type 1 agent. Suppose

---

[32] Similar expressions follow for agent 2.

[33] In Demski and Sappington [1984] and in Ma, Moore, and Turnbull [1988], this inequality is strict since they assume the agents are risk averse.

Agent 2

	type 1	type 2
type 1	(2,2) (4,4)	(2,1) (4,2)
type 2	(1,2) (2,4)	(1,1) (2,2)

Agent 1 (label to the left of the table, aligned between type 1 and type 2)

FIGURE 4.3

Agent 2

	type 1	type 2	F
type 1	(2,2) (4,4)	(2,1) (4,2)	(1,2) (2,3.5)
type 2	(1,2) (2,4)	(1,1) (2,2)	(1,1) (2,0)
F	(1,2) (3,5.2)	(1,1) (0,2)	(2,2) (0,0)

FIGURE 4.4

that this is not a good outcome from the point of view of the principal, i.e. the principal is strictly worse off[34] at this equilibrium relative to the truthful one. As shown by Ma, Moore and Turnbull [1988], it is possible to augment the direct mechanism to eliminate such an equilibrium. Here, since we have made simplifying assumptions, it is possible to construct a particularly simple augmentation.

The augmentation consists of giving each agent an additional strategy, which in effect allows an agent to say that the other agent is not playing the truthful strategy. Consider the following indirect mechanism (Figure 4.4).

---

[34] Demski and Sappington [1984] show this to be the case in the fully specified model. With risk averse agents, they show that under the optimal contract, both agents can be *strictly* better off in such an equilibrium.

The additional strategy for each agent, labelled F for 'fink', allows the principal to eliminate the unwanted equilibrium. Consider the position of agent 2. If he is of type 1, then the expected payoff from finking when 1 is truthful is 1.4, while that from truth is 2. On the other hand, if agent 1 always claims to be type 1, then the expected payoff from finking is 2.5 as opposed to 2 from truth. Thus the unwanted equilibrium has been eliminated. The outcomes when agents fink have been set up so that it never pays to fink if the other agent says type 2. It is easy to verify that both agents finking is not an equilibrium (since anything else is better), and that in fact the unique equilibrium is the truthful one.

E. Public goods: private values and common values

Laffont and Maskin [1982] were, to our knowledge, the first to raise the issue of multiple equilibria in public goods mechanisms. Much of the early literature on public good provision with asymmetric information focused on dominant strategy implementation (e.g. Green and Laffont [1987]), where multiple equilibria are not consequential. However, the papers of Arrow [1979] and d'Aspremont and Gerard-Varet [1979] and parts of Laffont and Maskin [1979] and Green and Laffont [1979] introduced the Bayesian approach to this class of problems.[35] More recently, Guth and Hellwig [1986], Rob [1988], Postlewaite and Mailath [1989], Ledyard and Palfrey [1989] and d'Aspremont, Cremer, and Gerard-Varet [1990] have extended these results in several ways. In this section, we investigate the extent of the multiple equilibrium problem when players' preferences may depend upon other players' information.

1. Private values

We study the simplest possible model of public good provision. In this model, there are 2 agents, and a single indivisible unit of the public good can be produced. We assume that the cost of provision is zero. The model is easily extended to many agents and to include a cost of production. Agent i has a privately known valuation of the public

---

[35] The maximin (see Green and Laffont [1979] and Thomson [1979]) and Nash equilibrium (Groves and Ledyard [1977, 1979] and Walker [1980]) approaches have also been investigated.

good, say v_i, with $v_i \in [-1, 1]$, and suppose v_1 and v_2 are drawn independently from the same distribution, say G.

It is clearly optimal to produce the public good if and only if $v_1 + v_2 \geq 0$. The simplest mechanism we might consider to implement this allocation rule is the following. Each agent reports a valuation in $[-1, 1]$. If the sum of the valuations exceeds 0, the public good is produced. Otherwise, there is no provision.

Consider the problem facing agent 1. If his true valuation is v_1 and agent 2 is reporting truthfully, the expected payoff from reporting $\hat{v}_1$ is

$$\int_{-\hat{v}_1}^{1} v_1 \, dG(v_2).$$

It is straightforward to see that as long as $G' > 0$, this is maximized by setting $\hat{v}_1 = -1$ if $v_1 < 0$ and by setting $\hat{v}_1 = 1$ if $v_1 \geq 0$. This mechanism thus cannot implement the desired allocation rule, since the public good will be produced whenever either agent has a non-negative valuation. In fact, it can be shown that if the allocation space consists only of whether or not the public good is produced, there is no mechanism which will yield the desired allocation rule.

Positive results can be obtained if we assume that there is also a transferable good, say money. Let y^i be a payment made by agent i. Preferences over the public good and over money are given by:

$$U^i(v_i, y^i) = \begin{cases} v_i + y^i & \text{if the public good is produced} \\ y^i & \text{otherwise} \end{cases}$$

A mechanism maps $v = (v_1, v_2)$ into a vector (d, y^1, y^2) where d, the public good decision, is either 1 or 0 and y^1 and y^2 are the payments made to or received by the two agents. Feasibility (sometimes called budget balancing) requires $y^1 + y^2 = 0$, so that allocation rule does not use any excess resources and also does not waste resources. A mechanism is *successful* if, for all v, $d = 1$ if and only if $v_1 + v_2 \geq 0$. Therefore, feasible successful mechanisms are *ex-post* efficient.

The contribution of d'Aspremont and Gerard-Varet (1979) and Arrow (1979) was the discovery of a feasible, successful, and incentive compatible allocation rule for a class of public good problems. For the problem at hand, consider the following mechanism. Each agent reports a value in $[-1, 1]$, say $\hat{v}_1$ and $\hat{v}_2$. Outcomes are given by:

$$d(\hat{v}_1, \hat{v}_2) = 1 \text{ if and only if } \hat{v}_1 + \hat{v}_2 \geq 0,$$

$$y^1(\hat{v}_1, \hat{v}_2) = \int_{v_2} v_2 d(\hat{v}_1, v_2) dG(v_2) - \int_{v_1} v_1 d(v_1, \hat{v}_2) dG(v_1)$$

$$y^2(\hat{v}_1, \hat{v}_2) = \int_{v_1} v_1 d(v_1, \hat{v}_2) dG(v_1) - \int_{v_2} v_2 d(\hat{v}_1, v_2) dG(v_2).$$

It is easy to check that for any set of reports, this (direct) mechanism is balanced. To check incentive compatibility, consider the problem facing agent 1 when agent 2 is reporting truthfully. By reporting $\hat{v}_1$ when his valuation is v_1, his expected payoff is

$$\int_{v_2} [v_1 d(\hat{v}_1, v_2) + y^1(\hat{v}_1, v_2)] dG(v_2).$$

On examining the expression for y^1, it can be seen that the second term is independent of $\hat{v}_1$ and the first term is independent of $\hat{v}_2$. Thus the best choice of $\hat{v}_1$ for agent 1 is determined by maximizing

$$v_1 [1 - G(-\hat{v}_1)] + \int_{v_2 \geq -\hat{v}_1} v_2 dG(v_2).$$

Assuming differentiability of G, the first order condition requires $v_1 G'(\hat{v}_1) = \hat{v}_1 G'(\hat{v}_1)$, which implies $\hat{v}_1 = v_1$. A similar argument applies to agent 2, so that truth is an equilibrium to the mechanism. Thus this direct mechanism achieves the desired allocation rule as a Bayesian equilibrium.

The possibility of multiple equilibrium is suggested by the following observation of Laffont and Maskin [1982]. A natural alternative equilibrium is one where some traders overstate their true preferences and some traders understate their true preferences.

Let $I = 2$ and $G_1(\cdot) = G_2(\cdot) = \dfrac{\theta + 1}{2}$ on $[-1, 1]$. Define, for $k > 1$ a k-strategy given by:

$$\hat{v}_1(v_1) = -1 \qquad \text{if} \qquad kv_1 < -1$$

$$= 1 \qquad \text{if} \qquad kv_1 > 1$$

$$= kv_1 \qquad \text{if} \qquad kv_1 \in [-1, 1]$$

$$\hat{v}_2(v_2) = \frac{1}{k} v_2$$

The optimization problem for type v_1 of agent 1, given that 2 follows $\hat{v}_2(\cdot)$ is

$$\max_{\hat{v}_1} \int_{\hat{v}_1 + \hat{v}_2(v_2) \geq 0} v_1 \, dG(v_2) + \int_{\hat{v}_1 + v_2 \geq 0} v_2 \, dG(v_2)$$

But since G is uniform on $[-1, 1]$ and $\hat{v}_2(v_2) = \frac{1}{k} v_2$, this reduces to:

$$\max_{\hat{v}_1} \frac{1}{kv_1} (1 + k\hat{v}_1) + \tfrac{1}{2}(1 - \hat{v}_1^2)$$

So, assuming that second order conditions are satisfied and as long as $k\hat{v}_1 \in [-1, 1]$, $\hat{v}_1(\cdot)$ is a best response to $\hat{v}_2(\cdot)$. But if $k\hat{v}_1 > 1$ $\left(\text{i.e. } v \geq \dfrac{1}{k^2} \right)$ then the best response to $\hat{v}_2(\cdot)$ is to bid $\dfrac{1}{k}$, the maximum bid by agent 2 under $\hat{v}_2(\cdot)$. Therefore, the strategy profile proposed by Laffont and Maskin [1982] is not an equilibrium.

The above strategy fails because the support of $v_1(\cdot)$ is different from the support of $v_2(\cdot)$. Modifying agent 1's strategy so that the supports match also fails. Suppose agent 1 were to use his best response as derived above.

Turning to agent 2, his objective conditioned on being type v_2 is:

$$\max_{\hat{v}_2} \int_{\hat{v}_2 + \hat{v}_1(v_1) \geq 0} v_2 \, dG(v_1) + \int_{\hat{v}_2 + v_1 \geq 0} v_1 \, dG(v_1)$$

The second term is simply $\tfrac{1}{4}(1 - \hat{v}_2^2)$. The first term is discontinuous in $\hat{v}_2$. Given agent 1's revised strategy, it equals:

$$0 \qquad\qquad \text{if } \hat{v}_2 < -\frac{1}{k}$$

$$\tfrac{1}{2} v_2 \left(1 - \frac{\hat{v}_2}{k} \right) \qquad \text{if } \hat{v}_2 \in \left(\frac{-1}{k}, \frac{1}{k} \right)$$

$$v_2 \qquad \text{if } \hat{v}_2 \geq \frac{1}{k}$$

This discontinuity in agent 2's payoff means that when agent 2 is type $v_2 = -1$ (or close to -1), then there exists small $\varepsilon > 0$ so that he is better off announcing $\dfrac{-1}{k} - \varepsilon$ instead of $\dfrac{-1}{k}$, since doing so generates a small decrease in the second term but a large increase in the first term. This announcement guarantees him that the public good will not be produced.

2. Revision effects and common values

The possibility of multiple equilibrium arises with different specifications of preferences. The particular version we consider is the 'revision effects' model, first studied by Myerson [1981] in the context of an auction problem. This departs from the private values model by assuming that each individual's utility function depends on the types of all agents in an additively separable way. Again, we consider only the two person case here.

Specifically, given a vector of 'interim valuations', (v_1, v_2), each agent revises his valuation, which generates a vector of 'revised valuations', (γ_1, γ_2), given by

$$\gamma_1 = \alpha_1 v_1 + (1 - \alpha_1) v_2$$

$$\gamma_2 = \alpha_2 v_2 + (1 - \alpha_2) v_1$$

Individual utility functions are then defined as before, with γ_i replacing v_i. Recall that the private values model sets $\alpha_1 = \alpha_2 = 1$. We note that if $\alpha_1 = \alpha_2 = 1/2$, then $\gamma_1 = \gamma_2$ for all (v_1, v_2). We refer to this as the *pure common value* case.

A particularly simple case to analyze here is the one in which $\alpha_1 = \alpha_2$. In his case, successfulness of a mechanism simply retains $d(v) = 1$ if and only if $\hat{v}_1 + \hat{v}_2 > 0$, as before.[36] The transfer rule needs a minor modification:

---

[36] If $\alpha_1 \neq \alpha_2$, then the optimal mechanism would have to be asymmetric. This would complicate the analysis somewhat, but the essential points would still hold. A more significant extension would have (α_1, α_2) as private information, and what happens then is an open question.

$$y^1(\hat{v}) = \int\limits_{v_2 \geq -\hat{v}_1} (2\alpha - 1)v_2 dG(v_2) - \int\limits_{v_1 \geq \hat{v}_2} (2\alpha - 1)v_1 dG(v_1)$$

A similar expression[37] holds for agent 2.

In the special case in which $\alpha = \frac{1}{2}$, note that no transfers are ever made, i.e. $y^i(v_1, v_2) = 0$ for all v_1 and v_2. To check that truth is an equilibrium, note that the problem facing agent 1 is to choose $\hat{v}_1$ to maximize

$$(1/2)\int\limits_{v_2} [v_1 + v_2]d(\hat{v}_1, v_2)dG(v_2) = v_1[1 - G(-\hat{v}_1)]$$

$$+ \int\limits_{v_2 \geq -\hat{v}_1} v_2 dG(v_2).$$

This is exactly the same expression in the private values model with transfers, and the same argument shows that truth is an equilibrium. More generally, the maximization problems for the two agents are:

$$\max_{\hat{v}_1} \int\limits_{v_2 \geq \sigma_2^{-1}(\hat{v}_1)} [\alpha v_1 + (1 - \alpha)v_2]dG(v_2) + \int\limits_{v_2 \geq -\hat{v}_1} (2\alpha - 1)v_2 dG(v_2)$$

$$\max_{\hat{v}_2} \int\limits_{v_1 \geq \sigma_1^{-1}(\hat{v}_2)} [\alpha v_2 + (1 - \alpha)v_1]dG(v_1) + \int\limits_{v_1 \geq -\hat{v}_2} (2\alpha - 1)v_1 dG(v_1)$$

Again, it can be checked that truthtelling is an equilibrium. Next, consider the case when G is uniform in $[-1, 1]$, as before. With no rezision effects ($\alpha = 1$) recall that the second term is increasing from -1 to 0 and decreasing from 0 to 1. The left-hand term may be increasing or decreasing, depending on $\hat{v}_2(\cdot)$ or $\hat{v}_1(\cdot)$.

In particular, if $\hat{v}_2(\cdot)$ is constant and equal tr either 1 everywhere or -1 everywhere, then the left-hand term is flat in $\hat{v}_1$. In other words, if either agent always claims that he is the highest type then the other agent cannot affect the public decision, and similarly if either

---

[37] If $I > 2$, then let the revision effect be $\dfrac{1 - \alpha}{I - 1}$ and the coefficient on the integral of $t(\cdot)$ be $\dfrac{2\alpha - 1}{I - 1}$. The remainder of the analysis follows in a straightforward fashion.

agent always claims to be the low type. Therefore it is a simple exercise to check if such pooling strategies can be sustained in equilibrium, since it only depends on the right-hand term. Always claiming to be the highest type is an equilibrium as long as

$$\frac{\partial \int_{-\hat{v}_1}^{1} (2\alpha - 1)v_2 dG(v_2)}{\partial \hat{v}_1} \geq 0 \text{ at } \hat{v}_1 = 1$$

and

$$\frac{\partial \int_{-\hat{v}_2}^{1} (2\alpha - 1)v_1 dG(v_1)}{\partial \hat{v}_2} \geq 0 \text{ at } \hat{v}_2 = 1$$

These derivatives equal $-\frac{1}{2}(2\alpha - 1)\hat{v}_1$ and $-\frac{1}{2}(2\alpha - 1)\hat{v}_2$ respectively, so $\{\hat{v}_1(v_1) = 1; \hat{v}_2(v_2) = 1\}$ is an equilibrium if and only if $\alpha \leq \frac{1}{2}$. Similarly, $\{\hat{v}_1(v_1) = -1; \hat{v}_2(v_2) = -1\}$ is an equilibrium if and only if $\alpha \leq \frac{1}{2}$.

One may also notice that this argument only depended on the *derivative* of the right-hand term (the transfer term), not on its exact functional form. Therefore, since incentive compatibility *requires* the derivative of the transfer term to equal $\frac{1}{2}(2\alpha - 1)\hat{v}_1$ in any successful mechanism, this multiple equilibrium problem will arise in any successful direct game.

What is the interpretation of α? If $\alpha = \frac{1}{2}$, the natural interpretation is that we have a purely common value problem. Everyone values the public good the same. Thus there is an important information externality by telling the truth. Here there are essentially two kinds of equilibria, much like we see in a typical coordination game. In one, everyone tells the truth, the efficient outcome is selected, *and there are no transfers* in equilibrium. In the other kind of equilibrium, no information is transmitted — the good is either always produced or never produced — and there are still no transfers.

Another interpretation of $\alpha \leq \frac{1}{2}$ is that v_i incorporates a private value term, but both agents are altruistic and have a utility function that is a linear combination of their value and the other agent's value. Yet a third interpretation is that v_i incorporates some private value components and some common value components. With only 2 agents this would suggest that $\alpha > \frac{1}{2}$. With more than 2 agents, however, this explanation could be consistent with $\alpha \leq \frac{1}{2}$. In particular, suppose for example that there are I agents, and utility functions are

$$U_i(d, t_i, v) = \frac{2}{3}\left[\frac{\sum\limits_{j=1}^{I} v_j}{I}\right] + \frac{1}{3}v_i + t_i \qquad \text{if } d = 1$$

$$= t_i \qquad \text{if } d = 0$$

We conjecture that for sufficiently large I, pooling equilibria arise in a generalized d'Aspremont and Gerard-Varet mechanism, because as I becomes large, the model approximates the two agent case[38] with $\alpha = \frac{1}{3}$.

This revision effects model is not directly covered by Theorem 4.1. However, the Theorem extends to this case, as we show below. Thus, all multiplicity problems in this class of models can be resolved by using indirect mechanisms.

To complete this section, we indicate how Theorem 4.1 can be extended to include revision effects. More generally, suppose that the *ex-post* utility function of agent i is $U^i(a, t_i) + \phi^i(a, t_{-i}) + y^i$. Then, given an allocation rule $(p, y), p : T \to \mathbb{P}(A), y : T \to \mathbb{R}^1$, define $P^i(t_i)$ as before, but redefine the expected transfer to be

$$Y^i(t_i) = \int\int \phi^i(a, t_{-i}) [dp(t) - dP^i(t_i)] dG(t_i) + \int y^i(t) dG(t_{-i})$$

Then, it is straightforward to check that the same argument as in the proof of Theorem 4.1 essentially implements any interim efficient allocation rule with these redefined reduced form allocation rules. The fact that the dependence of preferences on the types of the other agents appears in a separable manner allows us to include it into the expected transfer, and the common-value part of the preferences can thus be treated separately from the private values part.

---

[38] Of course as I increases, the distribution of the average valuation of the agents

$$\left(\frac{\sum\limits_{j=1}^{I} v_j}{I}\right)$$

changes, which makes it difficult to analytically solve for the pooling conditions.

Summary

In this chapter, we have seen that the multiple equilibrium problem arises in many of the models frequently studied by economists. In addition, we have seen that in some important cases, certain desirable allocations *cannot* be implemented, though positive results are available in some settings. The characterization results of the previous section and the results presented here, all show that the problems are easily solved in the transferable utility case.

5. PREPLAY COMMUNICATION AND RENEGOTIATION

A. Issues of commitment and control

In the discussion so far, we have implicitly made two assumptions about what might be called 'the power of the planner'. First, we assume that the planner has sufficient *control* over the message space that the only possibility of communication by the agents involves sending messages to the planner from a prespecified list of allowable messages. A strategy for an agent is then limited to be a function which specifies a message, or report, as a function of the agent's type, and this message *must be an element of the agent's message space*. This implies, in particular, that we have ignored the possibility that agents may be able to communicate with each other in ways other than that specified by the mechanism.

A second, related, assumption is that the planner has *committed* to a prespecified outcome function. Thus, for example, the planner may be able to commit to outcomes which are known to be bad outcomes for all parties. This type of construction is particularly evident in the characterizations involving NEI information structures, where incentive compatibility is achieved by forcing contracts which threaten to destroy the entire endowment if particular profiles of messages are sent by the agents. Thus, 'good' equilibrium behavior is sometimes elicited by the planner by making outrageous threats to the agents if their reporting strategies are not the equilibrium strategies. Thus, *after* the mechanism is played, there might be reasons why everyone (perhaps even the planner) would want to change the outcome that is dictated by the mechanism. We will refer to this problem as *ex-post commit-*

ment.[39] While this is the most obvious sort of commitment that can be identified as questionable, there are other, more subtle issues relating to commitment to the outcome function. For example, in many contracting applications, we think of the planner simply as a mediator who is hired by the contracting parties as an 'assistant' to merely help carry out the wishes of the contracting parties. Thus, if the contracting parties mutually (unanimously) agree to change the mechanism *even before they have played it*, they can simply give this assistant a different set of instructions for what messages he should be collecting from everyone and how he should translate these messages into outcomes. Because this problem arises *before* the mechanism has been played, but presumably after players have observed their type, we call this the problem of *interim commitment*.

This chapter relaxes these assumptions of perfect control over the message space and perfect commitment to an outcome function. First, we examine the possibility of implementing allocation rules when the agents cannot be prevented from communicating with each other before they play the mechanism. Second, we explore the implications of allowing the players to renegotiate the entire mechanism (both the outcome function and the message space) at the interim stage, thereby relaxing the assumption of interim commitment.

B. Implementation with preplay communication

It is widely known that the Nash equilibrium outcomes of a game may be expanded by allowing players to engage in preplay communication before selecting their strategies. The subsequent, more complicated multistage game, may allow players to do two things that they couldn't do without communication: *coordinate their strategies* and *transmit useful information*. The following two examples illustrate these two possibilities.

---

[39] This type of problem has been the focus of a number of recent papers which investigate the implications of the players being able to *renegotiate* to better outcomes if a profile of messages happens to be sent which would lead to an undesirable outcome under the outcome function of the mechanism. See for example, Dewatripont [1989], Aghion, Dewatripont, and Rey [1989], Baron and Besanko [1987], Rubinstein and Wolinsky [1991], Green and Laffont [1987a, 1987b], Laffont and Tirole [1990], Hart and Tirole [1987], Fudenberg and Tirole [1990] and the references they cite.

Example 5.1 (The Battle of the Sexes Game)

This is a well-known two person bimatrix game where players A and B have payoffs given by the following payoff matrix:

		PLAYER B	
		L	R
PLAYER A	U	2, 1	0, 0
	D	0, 0	1, 2

There are three Nash equilibria to this game, two in pure strategies (yielding utility outcomes (2, 1) and (1, 2), respectively) and one in mixed strategies (yielding expected utilities (2/3, 2/3)). Suppose, before choosing their actions in this game, the players may first (simultaneously) announce to their opponent what strategy they plan to do. The second stage of this expanded game is then simply the original game. There are now many more equilibrium outcomes to this 2-stage game compared to the original 1-stage game, even though we have not added any 'payoff relevant' strategies for the players. For example, there is the equilibrium identified by Farrell and Saloner (1985), in which, in the second stage, the players carry out their announcements if the announcement profiles were ('U', 'L') or ('D', 'R'), and otherwise play the mixed strategy equilibrium. Given that these will be the strategies in the second stage, the game in the first stage is essentially converted to the following normal form game:

		PLAYER B	
		'L'	'R'
PLAYER A	'U'	2, 1	5/9, 5/9
	'D'	5/9, 5/9	1, 2

This has a mixed strategy Nash equilibrium where player A announces 'D' with probability 4/17 and player B announces 'L' with probability 4/17. This completes the description of one equilibrium of the two stage game. The resulting (expected) payoffs in this equilibrium are (137/153, 137/153), which Pareto dominate the mixed strategy equilibrium of the 1-stage game without communication.

A second equilibrium of interest in the 2-stage game has both players choosing their announcements completely randomly (i.e. each announcement has probability 1/2), and then subsequently playing (U, L) if the announcements were ('U', 'L') or ('D', 'R') and playing

(D, R) if the announcements were ('U', 'R') or ('D', 'L'). This produces expected payoffs $(3/2, 3/2)$, which is the highest possible symmetric expected payoff possible.

Both of these equilibria solve a complicated coordination problem associated with the original game, where the only equilibrium that produces symmetric payoffs is the 'bad' mixed strategy equilibrium. Thus no information is transmitted in the preplay communication, but the communication allows the players to correlate their play in the second stage. In the next example, there is no coordination problem, but the players have private information that they would like to share with the other player.

Example 5.2 The contribution game (Palfrey and Rosenthal [1991])

There are I players. Each player has two possible actions: 'Contribute' and 'Not Contribute'. If all players contribute, then each player gets a payoff of 1 (the public good is produced). If fewer than I players contribute, those who contributed get 0 (i.e. the public good is not produced and their contribution is wasted) and those who did not contribute (say, player i) get a payoff of c_i, which is private information. Each c_i is independently drawn from the interval $[0, W]$, $W > 1$, according to a commonly known distribution $F(c)$ which is continuous and strictly increasing on $[0, W]$. Without communication, all equilibria have the property that there are some realizations of types such that $c_i < 1$ for all i and the public good is not produced. In fact for many distributions, for example if F is Uniform on $[0, W]$, the only equilibrium is for no one to ever contribute. However, suppose that the players can simultaneously announce to each other whether their c_i is above or below 1. Then there is an equilibrium of the resulting 2-stage game with all players announcing honestly in the first stage. In the second stage, everyone adopts the strategy of contributing if and only if everyone announced their c_i was less than or equal to 1.

Notice that in the second example, the problem is not one of coordinating among multiple equilibria, as was the case in the first problem. Rather, the problem is one of information transmission. Allowing communication fully solves the information transmission problem, in the sense that the outcome is *ex-post* Pareto optimal, and is the unique *ex-post* Pareto optimum that satisfies an *ex-post* individual rationality constraint. In many ways, the problem of mechanism design with incomplete information, and the problem of Bayesian implementation, involves the design of communication devices (using a mediator) which will allow for information to be adequately transmitted and strategies to be properly coordinated. This is a point made in the work of Myerson [1986], Forges [1986] and others. The last example shows how sometimes even unmediated communication, coupled with a simple 'plannerless' mechanism can achieve efficient outcomes.[40]

However, the implementation approach requires more that the existence of an equilibrium that produces the desired allocation rule. The desired allocation must be the unique equilibrium outcome of the mechanism. In example 5.2, the 'good' communication equilibrium is not the unique equilibrium outcome. There are other communication equilibria, including the outcome where no one ever contributes. Thus we see that adding communication can expand the set of equilibria, but never contracts the set of equilibria, thereby compounding the uniqueness problem.

One might argue that in example 5.2, there is a 'natural' communication equilibrium, namely the one where everyone is honest and the efficient outcome results. While this may be true in this example, the next two examples show how preplay communication can undermine the implementation of *efficient*, allocation rules, in the sense that without communication, an efficient allocation rule is the unique Bayesian equilibrium (in fact *dominant strategy equilibrium*) to the mechanism, but allowing communication introduces other Bayesian equilibria.

Example 5.3 (Adapted from Palfrey [1990])

There are three agents and two alternatives, a and b. Agents 1 and 2 have two possible types, $T^1 = \{t_1, t_2\}$ and $T^2 = \{s_1, s_2\}$, while agent 3 has only one type. The utilities of the agents for a and b as a function of the vector of types are as shown in Table 5.1.

Suppose that the types of agents 1 and 2 are independent, and that $G(t_1) = 0.5$, $G(s_1) = 0.4$. Consider the allocation rule (Figure 5.1).

---

[40] A related point is made in the work of Banks and Calvert [1990], Ledyard and Palfrey [1989], Forges [1988], and Matthews and Postlewaite [1989].

TABLE 5.1

	(t_1, s_1)			(t_1, s_2)			(t_2, s_1)			(t_2, s_2)		
	1	2	3	1	2	3	1	2	3	1	2	3
a	0	0	1	1	1	1	1	1	0	0	0	0
b	1	1	0	0	0	0	0	0	1	1	1	1

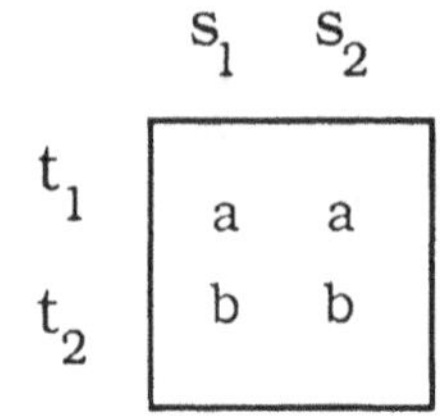

FIGURE 5.1

It can be checked that this allocation rule is incentive compatible, and in fact is the unique equilibrium outcome to the above direct mechanism, so it is implementable. Note that the allocation rule chooses in favor of agent 3 at the expense of agents 1 and 2.

Now, suppose that agents 1 and 2 could communicate with each other at the interim stage before playing this mechanism. Suppose, in particular, that agent 2 simply informs agent 1 about his type. Of course, agent 1 need not listen to agent 2, but could ignore the message and play the mechanism as before. However, the possibility of communication gives rise to a different equilibrium. In this equilibrium, agent 1 reports truthfully if agent 2 says he is of type s_2, and reports the opposite of his true type otherwise. It can be checked that given this strategy of agent 1, agent 2 should always communicate his type truthfully to agent 1 and also play truthfully in the mechanism. Given this strategy of agent 2, it can also be checked that agent 1's strategy is a best response. With these strategies, we get an equilibrium outcome with communication which differs from the unique equilibrium outcome without communication.

Example 5.4 The double auction with preplay communication (Matthews and Postlewaite [1989] and Farrell and Gibbons [1989])
Recall that in the double auction, there is one indivisible good, a buyer

and a seller. The buyer has a valuation, $v_b \in [0, 1]$, and the seller has a valuation $v_s \in [0, 1]$. Consider the 'buyer's bid' double auction, the special case of a k-double auction when $k = 1$. Riley and Zeckhauser [1983] show that the following strategies form an equilibrium. The seller reports his valuation truthfully, and the buyer chooses a bid b to maximize $(v_b - b)G(b)$. The outcome from this equilibrium is also the best possible outcome from the point of view of the buyer. Matthews and Postlewaite [1989] show that with communication, one can get an equilibrium outcome which is the best possible outcome for the *seller*! Further, this outcome with communication is not an equilibrium in the buyer's bid double auction without communication.

Summarizing these examples, the possibility of preplay communication has a good side and a bad side. On the one hand, it may offer players opportunities to coordinate and transmit information in relatively simple mechanisms and achieve better equilibrium outcomes. On the bad side, the fact that pre-play communication can expand the set of possible equilibrium outcomes poses problems for the theory of implementation developed in the previous chapters. In many applications, it is difficult to argue either that a specified mechanism captures all possible communication possibilities between agents, or that further communication can be precluded. This raises the following question. If we cannot control communication, is it still possible to design mechanisms which produce only desirable equilibrium outcomes? If an allocation rule can be implemented by a mechanism even when additional communication is allowed, we say that the allocation rule can be implemented by a communication proof mechanism.

This problem of *communication proof implementation* has been studied by Palfrey and Srivastava [1991b] in environments with transferable utility, independent types, and a limited form of dependent values. They show that in this class of environments, there is a close connection between efficiency and communication proof implementation. Specifically, interim efficient allocations can be implemented by a very simple kind of communication proof mechanism. We turn next to an exposition of this analysis.

To keep the exposition simple, we will examine the private-values, independent types model with transferable utility introduced in the last section, where A is a finite set of alternatives, $A = \{a_1, \ldots, a_m\}$, and $\mathbb{P}(A)$ is the set of all probability measures on A. If agent i is of type t_i, we write his *ex-post* utility as $U^i(a, t_i) + y^i$ where y^i is the transfer to i. An allocation rule consists of a pair of functions, $p : T \to \mathbb{P}(A)$ and $y : T \to \mathbb{R}^I$. For each vector of types, the first function gives the probability of a particular alternative being chosen while the second specifies transfers between agents.

The *reduced form* allocation rule is given by a collection of $\{P^i(\cdot),$ $Y^i(\cdot)\}$ pairs of functions, one for each i. Recall that $P^i(t_i)$ is the marginal distribution on A as a function of agent i's type, while $Y^i(t_i)$ is the expected transfer to i when of type t_i.

Theorem 4.1 in the previous chapter stated that if (p,y) is interim efficient then it is (essentially) implementable. Recall that the implementing mechanism was constructed as follows. The message space for each i was $M^i = T^i \times [0,1]$. If for all i, agent i reports $(t_i,0)$, then the outcome is $(p(t),y(t))$. If some agents report a positive number in the second component of their message space, then pick the agent reporting the lowest number, say i. The outcome is determined by the reduced form lottery, $P^i(t_i)$, the transfer to i is $Y^i(t_i) - \varepsilon$, and the other agents divide $-(Y^i(t_i) - \varepsilon)$.

Now, suppose that agents can communicate with each other before playing this mechanism. One problem we face is to define the possible ways by which this communication can take place. For example, one may wish to model the situation in which some communication is private while other communication is public. We model communication as follows. Suppose there are K stages of communication. Let M_k^i denote the set of messages agent i can send in stage k, and let N_k^i be the set of messages agent i can receive in stage k and let $M_k = M_k^1 \times \ldots \times M_k^I$. For each $k = 1, \ldots, K$, and for each i, the function $h_k^i : M_1 \times \ldots M_k \to N_k^i$ specifies a message that agent i receives at the end of stage k.[41] We assume that agents engage in this communication at the interim stage, i.e. after they have observed their own types. The collection $\{K,(M_1,N_1,h_1), \ldots, (M_K,N_K,h_K)\}$ is called a *communication procedure*. Finally, after the K stages of communication, agents play the original mechanism, (M, g), but now their strategies in (M, g) may be a function of the messages they sent

---

[41] One could let the message i receives in each stage k be a function of a random variable. One could also let i's set of allowable messages in stage k, M_k^i, be a function of the past messages sent by all players, and even a function of a random variable, as well. None of these complications affect the results described below.

and the messages they received, as well as their type.

Call the entire game, i.e. the K stages of communication followed by the mechanism (M,g) a K-stage communication extension of (M, g). In this game, a strategy for agent i consists of a set of functions.

$$\sigma_1^i: T^i \to M_1^i$$

$$\sigma_k^i: T^i \times N_1^i \times N_2^i \times \ldots \times N_{k-1}^i \times M_1^i \ldots \times M_{k-1}^i \to M_k^i \text{ for}$$

$$k = 2, \ldots, K,$$

and

$$\sigma^i: T^i \times N_1^i \times N_2^i \times \ldots \times N_K^i \times M_1^i \ldots \times M_K^i \to M^i.$$

Thus, at each stage, agents communicate, and after the communication stages are over, they play the original mechanism. The outcome function, g, is still a function of M alone. This embodies the notion that the communication is 'cheap-talk' and can affect outcomes only indirectly through the strategies of the agents.

An equilibrium is defined as before: given the strategies of the other players, each player's strategy is a best response. This response includes the consideration of all information they receive during the communication phase. That is, it must be a Bayesian equilibrium of the K-stage communication extension of (M, g).

Definition 5.1: (p, y) is *essentially implementable by a communication proof mechanism* if there exists a mechanism (M, g) which essentially implements (p, y), and such that if σ is an equilibrium to some K-stage communication extension of (M, g), then $W^i(t_i, \sigma) = V^i(t_i)$ for all i and t_i.

Theorem 5.1: Let (p, y) be interim efficient. Then (p, y) is essentially implementable by a communication proof mechanism.

Proof: Consider the mechanism constructed in the proof of Theorem 4.1. We already know this mechanism essentially implements (p, y), so suppose there is an equilibrium σ to some K-stage communication extension of this mechanism with $W^j(t_j, \sigma) \neq V^j(t_j)$ for some j and t_j. Since (p, y) is interim efficient, it must be the case that there exists some type t_i of some agent i such that $W^i(t_i, \sigma) < V^i(t_i)$. But then this type of agent i could modify his strategy by communicating in exactly the same way but reporting a small $\varepsilon > 0$ after the communication stage. Hence $W^i(t_i, \sigma) = V^i(t_i)$ for all i, t_i. ■ ■ ■

This result shows that in certain environments, it is possible to implement efficient allocations despite the possibility of pre-play communication. Palfrey and Srivastava [1991b] show how the result can be extended to allow for certain kinds of common values. However, there are as yet no general results with dependent types.

C. Renegotiation-proof implementation

The problem of renegotiation in the design of contracts and mechanisms has recently emerged as one of the critical practical issues in the theory of mechanism or contract design. The problem arises because it is often either impossible or very difficult for the interested parties to commit *not* to change the rules governing the mechanism or contract. Therefore, a mechanism simply provides a benchmark which the players might eventually end up agreeing to alter, in the event that a circumstance arises whereby a change in the rules would benefit all parties. This inability to commit in advance to a set of rules which cannot be altered can undermine the value of 'binding' contracts. While contracts are typically unilaterally binding on all parties who sign the agreement, many contracts can be replaced by a new one by unanimous consent. Thus, contracts which enforce desirable outcomes using 'threats' which are undesirable for all parties may fail to pass a common-sense credibility test, thereby undermining the incentive structure of the contract.

Thus it is clear that, above all, the problem of renegotiation in mechanism design is linked to *efficiency*. Contracts are renegotiated when an inefficient outcome occurs. Second, the problem of renegotiation is sensitive to assumptions about the *information* of the parties. All of the parties must be able to agree on a Pareto improvement since an alteration of the original contract requires unanimous consent. Thus, the parties must know, at the time[42] of renegotiation, that there are opportunities to change the contract for the benefit of

---

[42] The issue of exactly what is the *timing* of the renegotiation process will be addressed below.

everyone. Third, there is the question of *enforcement*: under what circumstances is it reasonable to assume that a third party exists which can enforce the original agreement? If the rules of the original mechanism can be enforced, then the renegotiation problem at the *ex-post* stage (after parties have played the 'message-sending' part of the mechanism, and have selected any private actions) is inconsequential. Nevertheless, the opportunity for renegotiation may still arise at the interim stage which occurs after the parties acquire information but before the mechanism is played out.

We call this *interim renegotiation*.[43] This problem was first addressed in a pair of closely related papers by Myerson [1983], Holmstrom and Myerson [1983], and Crawford [1985], and reformulated more recently in papers by Cramton and Palfrey [1989], Legros [1990], and Maskin and Tirole [1992]. The Holmstrom and Myerson paper introduced a notion of allocation rule *durability*, which attempted to capture the intuition that (in addition to incentive compatibility) at the interim stage it should never be possible for the original contract to be replaced with some other contract by a unanimous vote of all parties. That paper also introduced the idea of interim efficiency and demonstrated the existence of durable, interim efficient mechanisms for one class of environments. Nevertheless, they also showed via an example that durability requirements could undermine some interim efficient allocation rules in more general environments.

Legros [1990] has introduced a slightly different durability concept and demonstrates that use of the indirect mechanisms proposed in Palfrey and Srivastava [1991b] vastly expands the set of durable allocation rules. This reformulation shifts the focus of durability somewhat away from allocation rules and in the direction of the mechanisms themselves. Legros argues quite convincingly that this distinction, which tends to be blurred in analyses that rely heavily on the revelation principle, is essential in this case, since mechanisms with large message spaces (indirect mechanisms) may be less vulnerable to proposals for alternative mechanisms. This is an important insight.

---

[43] In particular, we will assume that the *ex-post* outcomes of a mechanism can be enforced by a third party. In this sense, our work is quite distinct from the work by Hart and Moore [1986], Green and Laffont [1987b], Aghion, Dewatripont, and Rey [1989], and Maskin and Moore [1988].

However, that paper shares the shortcoming of Holmstrom and Myerson by not explicitly modelling the renegotiation process itself. In particular, no agent 'proposes' an alternative contract. A contract fails to be durable if there exists an alternative at the interim stage that would be unanimously approved were it proposed by some exogenous 'third party'. This is related to the criticism of durability in Crawford [1985].

Cramton and Palfrey [1989] also study the durability problem and explore the implications of a proposed alternative *failing* to receive a unanimous vote to replace the original contract. In particular, they find that this possibility of 'learning from disagreement' may either expand or reduce the set of enforceable agreements that can be reached. Like the work on durability, the actual process of proposing alternative contracts (the 'renegotiation process') is not explicitly modeled.

The approach taken here is similar to the model of renegotiation in Maskin and Tirole [1992]. That paper reformulates the informed principal problem of Myerson [1983] as a non-cooperative game in which a principal proposes a contract, and an agent makes inferences about the principal's private information from the actual proposal.

The timing is the following (see Figure 5.2). At date 0, the parties agree to a mechanism before the principal observes his type, where a mechanism involves the principal making an announcement about his type, plus some nuisance messages by the agent and by the principal. At stage 1, the principal observes his private information. At stage 2P, the principal may propose an alternative mechanism to replace the original mechanism. At state 2A, the agent either agrees to the proposed mechanism or not. At state 3 the old mechanism is played out if no new mechanism was proposed at stage 2P or if a proposed mechanism was vetoed by the agent at stage 2A. If a new mechanism was proposed and approved in stage 2, then it is played out in stage 3. The outcomes are enforced in stage 4 (even if they are *ex-post* inefficient). An allocation rule is (strongly) renegotiation-proof if all equilibria of the continuation game in stages 2, 3 and 4 produce allocations that are interim-utility-equivalent to it. In other words, if an allocation rule is renegotiation proof, then there is a mechanism which produces that allocation rule in equilibrium with or without renegotiation. They prove that an allocation rule is renegotiation-proof if and only if it is interim efficient.

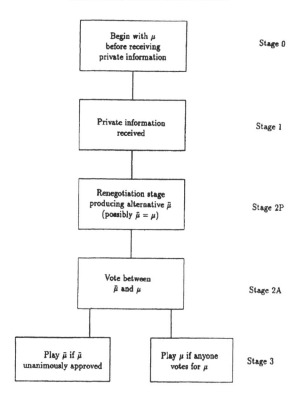

FIGURE 5.2

A natural next step is to investigate the extent to which these results may generalize when there are multiple informed parties. The rest of this section provides some initial results in this direction.

At least three interesting issues arise when multiple privately-informed parties are involved. First, the class of renegotiation procedures which need to be considered is much larger, since multilateral information leakage is possible. Both parties may have incentives to transmit or conceal information at the interim stage. Therefore, stage 2P is much more complicated, and a wider class of extensive forms must be considered. Second, information can also be leaked at stage 2A, in which all parties vote to approve a new mechanism that emerged from the proposal stage, 2P. Third, the multiple equilibrium problem is more severe. In contrast, Maskin–Tirole examine an environment in which, without renegotiation, any incentive compatible allocation rule can be uniquely implemented. The reason for this is that there is only one informed party. With multilateral asymmetric information, the multiple equilibrium problem is difficult to avoid, which is one of the central points of this monograph.

The general model is laid out below. We prove a result analogous to the efficiency characterization in Maskin–Tirole, under the assumptions of *independent types, reduced-form equivalence*,[44] and *economic environments*. In these settings, we prove that an allocation rule is renegotiation-proof implementable if and only if it is interim incentive efficient. We then discuss some difficulties in obtaining equivalence between interim efficiency and renegotiation-proofness in more general environments.

Let p be an allocation rule. Let μ be a mechanism to which the parties agree in stage 0, before receiving any private information. In stage 1, all parties observe their respective types. Stage 2P is the 'renegotiation' stage, during which players might follow some very complicated (or very simple) procedure for considering alternative mechanisms.

The following is one example of such a procedure. First, give party 1 an opportunity to propose a mechanism μ_1, after which party 2 would propose a mechanism μ_2. All parties would then vote between μ_1 and μ_2, the 'winner' being decided by majority rule, perhaps with a tiebreaking procedure. Call the winner μ_2^w. Then party 3 could propose μ_3 which would be voted against μ_2^w, the winner being denoted μ_3^w. This could continue, for an arbitrary, finite, number of rounds; for example, each party might get exactly one proposal turn. The (temporary) 'outcome' of this procedure would be the final winning proposal μ_I^w.

This example illustrates several requirements of what we will call a *renegotiation procedure*. In particular, there are *two* essential ingredients. First, it consists of something that resembles a communication procedure (where the messages communicated in this example have the interpretation of votes). Second, this communication procedure

---

[44] Reduced-form equivalence requires a separability between common-value components of private information and private value components. Thus, assuming private values will satisfy reduced form equivalence.

has 'outcomes' which are *mechanisms*. Other than this, we do not place any serious restrictions.[45]

The procedure may involve very complicated patterns of sending and receiving messages, 'secret' communication between subsets of parties, even 'mediated' communication through an uninterested party, or even through one of the interested parties. This can represent different voting procedures, multiple proposals can be put forth, and the organization of communication and proposal-making might be hierarchical, as it is in many real world organizations. The 'order' of communication and proposal making could be endogenous. We will assume there are no delay costs associated with the process, as was implicitly assumed in the analysis of communication proof mechanisms.

Denote by $\bar{\mu}$ the outcome of stage 2P. Stage 2A is very simple. Each agent votes for either μ of $\bar{\mu}$. If *all* agents vote for $\bar{\mu}$ then in stage 3 $\bar{\mu}$ is played. Otherwise μ is played in stage 3. Thus, in stage 2A, each party has veto power to reject $\bar{\mu}$. If $\bar{\mu}$ is rejected by anyone, then the original μ is played. Stages 3 and 4 proceed as in Figure 1.

This would seem to be the natural extension of the renegotiation model of Maskin–Tirole [1992] to the case of multilateral private information. It converts the standard mechanism design problem into a problem of multiple informed principals. With a single principal, it is sensible to look at only *one* renegotiation procedure, as in Maskin–Tirole. The single procedure they consider is for the principal to propose $\bar{\mu}$. Presumably one could show that this single procedure is without loss of generality when only one party is informed. That is, their characterization of renegotiation-proof mechanisms in terms of interim efficiency would be true for the more general class of renegotiation procedures we consider. Obviously with multiple informed parties such a simplification would not be appropriate.

Formally, let M_0 be a collection of mechanisms. For now, think of M_0 as the set of all mechanisms, as defined earlier. A *renegotiation procedure R* is a pair, (C, r) where $C = \{K, (M_1, N_1, h_1), \ldots, (M_K, N_K, h_K)\}$ is a communication procedure and $r : M_1^i \ldots \times M_K^i \rightarrow M_0$ is a proposal procedure.[46] That is, r is a rule which con-

---

[45] We do assume players have perfect recall.

[46] One could define the proposal procedure to allow the proposed mechanism to depend on a chance move as well.

verts the messages transmitted by all the agents into a 'proposed mechanism'.

A voting stage follows the last stage in R. Given some sequence of messages in c, say m, a proposal has resulted, denoted $\bar{\mu} = r(m)$. In the voting stage, each player may cast a vote either for μ or for $\bar{\mu}$. A strategy for player i in the voting stage is a function $v^i \colon T^i \times M^i_1 \ldots \times M^i_K \to \{Y, N\}$. The 'outcome' of the voting stage is a mechanism, either μ or $\bar{\mu}$. If anyone votes 'N', then the outcome is μ; if everyone votes 'Y' then the outcome is $\bar{\mu}$. That is, $\bar{\mu}$ replaces μ if and only if it is unanimously approved.

Finally, the selected mechanism is played (either μ or $\bar{\mu}$) and the outcome of the mechanism is enforced.[47] A 'strategy' for player i in the mechanism that is finally played can depend not only on i's type but also on the entire history of plays that i has observed in the renegotiation process and in the voting stage.

Thus, a *mechanism selection process* (μ, R) consists of a (status quo) mechanism μ, a renegotiation process $R = (C, r)$ which occurs at the interim stage (after all players have observed their types), followed by a voting stage, followed by the play of the mechanism chosen in the voting stage. This defines a multistage game. We call a strategy for i in this process, σ^i_R. Let $\Sigma^*_R(\mu)$ be the set of $\sigma_R = (\sigma^1_R, \ldots, \sigma^I_R)$ that are Bayesian equilibria to the game defined by the message selection process (μ, R) and let $X^*_R(\mu)$ be the set of equilibrium outcomes associated with $\Sigma^*_R(\mu)$, and let $V^*_R(\mu)$ be the interim utility allocations of $X^*_R(\mu)$.

Definition 5.2: μ is *renegotiation proof* if, for all renegotiation processes R, $V^*_R(\mu) = V^*(\mu)$.

Definition 5.3: (p, y) is *renegotiation proof implementable* (at the interim stage) if there exists μ such that $V^*_R(\mu) = V(p, y)$, for all renegotiation procedures R.

Theorem 5.2:   An allocation rule (p,y) is renegotiation proof if and only if it is interim efficient.

---

[47] That is, there is no further (*ex-post*) renegotiation after the mechanism has been played.

Proof: The proof of 'if' proceeds by showing that the mechanism μ constructed in Theorem 4.1 is renegotiation proof.

To see that μ is renegotiation proof, suppose not. Then, there is a K-stage renegotiation procedure and an equilibrium, say σ, producing an outcome q. Since (p, y) is interim efficient, either $V^i(q, t_i) = V^i(p, t_i)$ for all i and t_i or there exists $\varepsilon > 0$ such that for some i and some $t_i \in T^i$ $V^i(q, t_i) < V^i(p, t_i) - \varepsilon$. In the latter case, consider the following alternative strategy, $\downarrow$ for player i:

$$\downarrow_k = \sigma^i_k \text{ for all } k \le K$$

$$\downarrow_{K+1}(\tau_i, r_K) = \sigma^i_{K+1}(\tau_i, r_K) \text{ for all } r_K \text{ and all } \tau_i \neq t_i$$

$$\downarrow_{K+1}(t_i, r_K) = (t_i, \varepsilon/2) \text{ for all } r_K.$$

This alternative strategy improves i's interim payoff by at least $\varepsilon/2$ when i is of type t_i and does not affect his payoffs at any other type. Hence, σ is not an equilibrium.

To prove 'only if', suppose that p is not interim efficient. If p is not implementable, then we are done, so suppose p is implementable by some mechanism μ. We need to show that μ is not renegotiation proof. Since p is not interim efficient, there exists an alternative allocation rule, q, such that every type of every player receives as high interim utility from q and some type t_i of some agent i receives strictly higher interim utility. Consider the very simple renedotiation procedure R in which i can either propose the direct mechanism defined by q or 'pass'. There is an equilibrium to this mechanism selection procedure in which all types of player i propose q in the renegotiation stage, all types of all players vote 'Y' (i.e. for q) in the voting stage, and all players report honestly in the final stage. Thus m is not renegotiatiol proof. ∎ ∎ ∎

While this establishes a positive result, several open questions remain. As in the results on pre-play communication, we do not know if this theorem extends to the case of dependent types or when there is no divisible private good.

6. OTHER TOPICS

In the previous section, we discussed the Bayesian implementation problem when the agents could not be prevented from communicating prior to playing the mechanism, and also when the mechanism itself could be renegotiated. In this section, we briefly discuss two other kinds of extensions. The first of these is implementation with an alternative equilibrium concept, and the second has both an alternative equilibrium concept and a weaker implementation notion.

A. Implementation with refinements

We have seen (in Sections 3 and 4) that direct mechanisms generally will not solve the implementation problem, and that direct mechanisms can frequently help in this regard. However, there are interesting examples which show that Bayesian equilibrium by itself does not impose sufficient restrictions on behavior to eliminate undesirable equilibria even with in direct mechanisms. Recall Example 4.3, which illustrates this.

Example 4.3: $I = 3$, $A = \{a, b\}$, $T^i = \{t_a, t_b\}$ for all i. Types are independently drawn with $q^i(t_b) = q$ for all i and $q^2 > .5$. Preferences are as follows: type t_a strictly prefers a to b, while type t_b strictly prefers b to a. We normalize utility so that $U^i(a, t_a) = 1 > 0 = U^i(b, t_a)$ and $U^i(b, t_b) = 1 > 0 = U^i(a, t_b)$. Note that for every t, there is a unique majority winner at t. The allocation rule given below simply chooses the majority winner at each t. This allocation rule has many nice properties, and is the only *reasonable* allocation rule in that it is incentive compatible; it is *ex-ante* efficient, interim efficient, and *ex-post* efficient. It satisfies all of the Arrow criteria (including independence of irrelevant alternatives and nondictatorship) (Figure 6.1).

Remarkably, x is not implementable in Bayesian equilibrium: let $\alpha^i(t_i) = t_b$ for all i, so $x(\alpha(t)) = b$ for all t. It is easily shown that there do not exist i, y, and t_i which satisfy the inequalities required by Bayesian monotonicity. Consequently, in *any* game in which σ is a Bayesian equilibrium with $g(\sigma) = x$, σ_α is also a Bayesian equilibrium with $g(\sigma_\alpha) = x_\alpha$. This has severe welfare implications, as $x_\alpha \equiv b$ violates all the notions of efficiency.

However, it is clear from examining the allocation rule that at each t_i, each agent has a dominant strategy to tell the truth in the direct mechanism. To see this, consider agent 1 at type t_a. No matter what the reports of the other agents, reporting t_a is always at least as good as reporting t_b, and is sometimes strictly better. Thus, in the

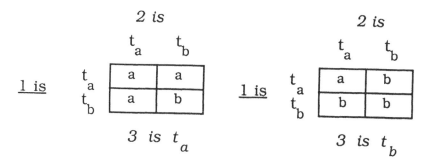

FIGURE 6.1

Bayesian equilibrium in which all agents always report type t_b, every agent is using a weakly dominated strategy (formally defined below).

In view of examples such as this one, Palfrey and Srivastava [1989b] examine the problem of implementation by *undominated Bayesian equilibrium*, which are Bayesian equilibria in which no agent uses a weakly dominated strategy.

Definition 6.1: Given a mechanism (M, g), the strategy $\sigma^i: T^i \to M^i$ is *weakly dominated for i* at t_i if there exists $\hat{\sigma}^i: T^i \to M^i$ such that

$$W^i(\hat{\sigma}^{-i}, \sigma_{t_i}^{-i}; M, g) \geq W^i(\sigma^i, \sigma_{t_i}^{-i}; M, g) \quad \text{for all } \sigma^{-i}, \text{ and}$$

$$W^i(\hat{\sigma}^i, \sigma_{t_i}^{-i}; M, g) > W^i(\sigma^i, \sigma_{t_i}^{-i}; M, g) \quad \text{for some } \sigma^{-i}.$$

Definition 6.2: Given a mechanism (M, g), a strategy profile σ is an *undominated Bayesian equilibrium* if σ is a Bayesian equilibrium and for all i and t_i, σ^i is nonweakly dominated for i at t_i.

Palfrey and Srivastava [1989b] describe necessary conditions and sufficient conditions for implementation by undominated Bayesian equilibrium. In a special class of environments, they show that the implementation problem can by completely solved in the sense that any incentive compatible allocation rule can be made the unique equilibrium outcome to a mechanism. These environments are defined by the following:

(i) Private values:
    $U^i(a, t) = U^i(a, t'_{-i}, t_i)$ for all t'_{-i}. (Hence denote $U^i(a, t_i) \equiv U^i(a, t)$.)

(ii) Value distinguished types:
      For all i, t_i, t_i', there exist a, b, $\in A$ with either
      $U^i(a, t_i) > U^i(b, t_i)$ and $U^i(a, t_i') \leq U^i(b, t_i')$ or
      $U^i(a, t_i) \geq U^i(b, t_i)$ and $U^i(a, t_i') < U^i(b, t_i')$
(iii) No complete indifference:
      For all i and t_i, there exist a, $b \in A$ with $U^i(a, t_i) > U^i(b, t_i)$
(iv) Existence of best and worst elements
      For all i and t_i, there exist b, $w \in A$ such that
      $U^i(b, t_i) \geq U^i(a, t_i)$ for all $a \in A$,   and
      $U^i(w, t_i) \leq U^i(a, t_i)$ for all $a \in A$.

Theorem 6.1 (Palfrey and Srivastava [1989b]): Assume conditions
(i)–(iv). Then x can be implemented in undominated Bayesian
equilibrium if and only if x is incentive compatible.

Note that this theorem is stated in terms of an allocation rule as
opposed to a social choice correspondence. The different assumptions
serve different purposes. Condition (iii) avoids certain trivialities.
One way to think about it is that in order to ensure that x_α is not
an equilibrium outcome for some α, the mechanism must give some
agent an incentive to deviate from σ_α. With complete indifference,
this is impossible. Condition (iv) is satisfied in many examples of
interest. The main import of conditions (i) and (ii) are that they make
it possible to create incentives *off the equilibrium path* which ensure
that lying is weakly dominated.

The result stated above is clearly a powerful one since in most studies
of strategic interaction with incomplete information, conditions
(i)–(iv) are satisfied. It is possible to characterize the conditions for
implementation without these assumptions; however, in general
domains, the conditions are difficult to verify and are much less
intuitive. A detailed discussion of these issues is contained in Palfrey
and Srivastava [1989b].

One feature of the construction employed in the proof of Theorem
6.1 is that it uses a technique which is controversial. In particular,
certain strategies are ruled out from being equilibria because they form
part of an infinite chain of successively dominated strategies. Thus,
strategy 1 is ruled out because it is weakly dominated by strategy 2,
2 by 3, and so on. Such techniques have been criticized by Jackson
[1989], who argues that faced with such a situation it is not clear that
an agent will not use a weakly dominated strategy and thus the

mechanism may not be entirely compatible with the solution concept. However, it appears that this unappealing type of construction (called 'tailchasing') is not essential to establishing this general result in most environments, given the results of Jackson, Palfrey and Srivastava [1990]. They show that in complete information settings, this technique is not needed for most social choice functions of interest. They identify a mild condition called 'chaining' that is sufficient to avoid the tailchasing constructions. With incomplete information, the main difference is that incentive compatibility conditions must be satisfied. Thus a reasonable conjecture is that with incomplete information, an interim version of the 'chaining' condition will lead directly to an extension of their results to the Bayesian case.

B. Virtual implementation

In all the discussion so far, we have focused on 'exact' (terminology of Abreu and Matsushima [1990b]) implementation in that we have required that the outcomes of the mechanism coincide exactly with the desired allocation rule(s). The theory of virtual implementation, introduced by Matsushima [1988] and Abreu and Sen [1991], relaxes this requirement. Virtual implementation only requires that the outcomes of the mechanism are approximations of the desired allocations. Both the above papers show that with complete information, virtual implementation in Nash equilibrium places no restrictions on implementable allocations.

Recently, Abreu and Matsushima [1990a] (see also Abreu and Matsushima [1992]) have developed the theory of virtual implementation in incomplete information environments, where the solution concept is the iterated elimination of strictly dominated strategies. We next discuss a simplified version of their analysis.

In the Abreu–Matsushima framework, there is a finite set of types, and the set of alternatives is $\mathbb{P}(A)$, where $\mathbb{P}(A)$ is the set of random allocations over A. The preferences of agent i given t are represented by a von Neumann–Morgenstern utility function, $U^i(a, t)$.

The equilibrium concept is as follows. Given a mechanism (M, g), with M^i finite for all i, and sets $H^i \subseteq M^i$ for all i, a strategy σ^i is strictly dominated for agent i with respect to $H = H^1 \times \ldots \times H^I$ if there exists $\hat{\sigma}^i \in H^i$ and t_i such that for all σ^{-i},

$$V^i(g(\hat{\sigma}^i, \sigma^{-i}), t_i) > V^i(g(\sigma), t_i) \text{ and}$$

$$V^i(g(\hat{\sigma}^i, \sigma^{-i}), t'_i) \geq V^i(g(\sigma), t'_i) \text{ for all } t'_i.$$

Thus, $\hat{\sigma}^i$ does at least as well as σ^i for every type of agent i and does strictly better at some t_i.

Let $Q_i(H)$ denote the set of strategies for i which are not strictly dominated for i with respect to H, and let $Q(H) = Q_1(H) \times \ldots \times Q_I(H)$. The set of iteratively strictly undominated strategies is then defined by letting $Q^0(M) = M$, $Q^k(M) = Q(Q^{k-1}(M))$ for $k \geq 1$, and finally, $Q^*(M) = \cap Q^k(M)$. Since M and T are finite sets, Q^* is nonempty and it contains all strategies that survive iterated elimination of strictly dominated strategies. A mechanism (M, g) *exactly implements* x *in iterative elimination of strictly dominated strategies* if $Q^*(M) = \{\sigma\}$ is a singleton and $g(\sigma) = x$. An allocation rule $x: T \to \mathbb{P}(A)$ is *virtually implementable in iterative elimination of strictly dominated strategies* if for every $\varepsilon > 0$, there exists $y: T \to \mathbb{P}(A)$ such that y is exactly implemented by some mechanism and $|y - x| < \varepsilon$.

A weak (but complicated) necessary condition for both virtual and exact implementation by this solution is *measurability*. A slightly oversimplified version of it says that if no one's interim preference relation changes between t and t', then the social choice function has to assign the same allocation at t and t' (i.e. $x(t)$ must equal $x(t')$). Unfortunately a complete formal definition of measurability is quite cumbersome. The reader is referred to Abreu and Matsushima [1990a, 1992] for the details. We summarize the basic ideas here.

Definition 6.3: Types t_i and t'_i are *interim value distinguished* if there exist random allocation rules $x: T^{-i} \to \mathbb{P}(A)$ and $z: T^{-i} \to \mathbb{P}(A)$ such that

$$V^i(x, t_i) > V^i(y, t_i) \text{ and } V^i(x, t'_i) \leq V^i(y, t'_i) \text{ or}$$

$$V^i(x, t_i) \geq V^i(y, t_i) \text{ and } V^i(x, t'_i) < V^i(y, t'_i)$$

First, define equivalence classes of types based on the definition of interim value distinction. This defines a partition of each player's set of types. An allocation rule is measurable with respect to these partitions if it assigns the same allocation rule to two type profiles that differ only within a single equivalence class of types of some player. Next, one proceeds by iterating this definition of equivalence classes using a definition of interim value distinction relative to measurable allocation rules. Eventually (since if the type sets are finite) one reaches a point at which this iterative procedure can go no further.[48] Denote this final equivalence class partition of types by T^*.

Definition 6.4: A social function x satisfies *measurability* if it is measurable with respect to T^*.

It is easy to see that measurability is a condition that will also be satisfied vacuously if all types of all players are interim valued distinguished over $\mathbb{P}(A)$. There are, of course, some examples in which types fail to be interim value distinguished relative to the set of (deterministic) allocation rules, A, but the types are value distinguished over $\mathbb{P}(A)$. The following example illustrates this possibility.

Suppose there are 2 players, the allocation space is $A = \{a, b, c\}$, the type sets are given by $T_1 = \{t_1, t_1'\}$ and $T_2 = \{t_2, t_2', \}$, players have private values, and types are independently distributed with $\text{Prob}(t_2) = \text{Prob}(t_1) = .9$. Let player 1 have preferences:

$$U(a, t_1) = 1 \qquad U(a, t_1') = 1$$

$$U(b, t_1) = .6 \qquad U(b, t_1') = .4$$

$$U(c, t_1) = 0 \qquad U(c, t_1') = 0$$

Thus, both types of player 1 have identical interim preference rankings between all pairs of nonrandom allocation rules that depend only on player 2's type. However, the types of player 1 have different interim rankings over the (constant) random allocation rules x and y where x is a 50/50 lottery between a and c and y is 'b for sure'.

In all applications of mechanism design theory we know of the environments are specified such that measurability is vacuously satisfied. For this reason, it is probably appropriately viewed as simply a technical assumption which imposes no meaningful economic restriction on allocation rules in most cases. As Abreu and Matsushima correctly point out, it is essentially just a generalization of a neutrality requirement that the labelling of player types is not redundant, i.e. if

---

[48] Notice that in the typical case in which all types are interim value distinguished over random allocation rules, this procedure does not create any equivalence classes, and so stops before any rounds of iteration. It is hard to imagine examples in which multiple rounds of iteration would be required.

a pair of player types is equivalent in the sense of not being interim value distinguished, then the social choice function should treat the two types equally.

Theorem 6.2: A social choice function is virtually implementable by iterated elimination of strictly dominated strategies if and only if it is incentive compatible and satisfies measurability.

The proof of this result is given in Abreu and Matsushima [1990a]. Subsequently, they have also shown that this theorem also extends to exact implementation when the solution concept is iterated elimination of *weakly* dominated strategies (see Abreu and Matsushima [1990a, 1992]). Instead of reproducing their proof here, we instead indicate the technique behind their construction by discussing an example adapted from Glazer and Rosenthal [1990]. The example assumes complete information and augments the allocation space to allow for transfers, but contains most of the intuition behind the construction.

Example 6.2: There are three agents, and either all are of type t_a or all are of type t_b. $A = \{\alpha, \beta\}$, and the preferences of the agents are as follows:

$$U^i(\alpha, t_a) = 1 > U^i(\beta, t_a) = 0$$

$$U^i(\beta, t_b) = 1 > U^i(\alpha, t_b) = 0$$

Thus type t_a strictly prefers α to β, while the preferences of type t_b are the opposite. Monetary transfers are also permitted, and preferences are assumed to be separable in the transfer and strictly increasing in the transfer, as in the discussion in Section 5. The allocation rule to be implemented is $x(t_a) = \beta$, $x(t_b) = \alpha$, so that the choice at each type is the worst element of the agents.[49]

Since there are three agents and complete information, incentive compatibility is satisfied. Further, measurability is trivially satisfied, so that the theorem states that this allocation rule is virtually implementable. Glazer and Rosenthal [1990] construct the following imple-

---

[49] Strictly speaking, the allocation rule specifies these alternatives and also zero transfers.

menting mechanism. Fix $\varepsilon > 0$, and let $\delta > 0$ denote the maximum permissible transfer to any agent. The message space of agent i is $M^i = \{t_a, t_b\}^{K+1}$ for an integer $K > 1/\gamma$, where $\gamma = \min\{\varepsilon/6, \delta/2\}$. Thus a message for an agent consists of a $K + 1$ vector of reports of types (of everybody since types are perfectly correlated). The first report of types determines the probability of α or β being chosen as follows. For each agent whose first report is t_a, add $\varepsilon/3$ to the probability of α being chosen, and similarly for each report of t_b. For $k > 1$, add $(1 - \varepsilon)/K$ probability to β if at the k'th announcement, two or more agents report t_a; otherwise, add $(1 - \varepsilon)/K$ to α. Finally, an agent pays γ/K for each $k > 1$ at which his report disagrees with the report of the other agents, and for each $k > 1$, an agent pays γ if both that agent and all others reported the same type at all earlier reports and the agent reports a different type at report k.

Given the outcome function, it is clear that reporting your type truthfully in the first report strictly dominates lying, since truthful reporting strictly increases the probability of getting the preferred outcome. Thus, we can eliminate all strategies which involve lying at the first stage. Now consider the second report. Given that the first report is truthful, truth in the second report strictly dominates lying since changing the report involves paying γ, and the fine is large enough relative to the gain from increasing the probability of the preferred outcome being chosen. Continuing in this way, the only strategy which survives iterated elimination of strictly dominated strategies is the truthful report, in which case the outcome at t_a is β with probability $(1 - \varepsilon)$, and that at t_b is α with probability $(1 - \varepsilon)$. Further, no fines are paid in equilibrium, so this mechanism virtually implements the desired allocation rule.

At least two aspects of this kind of construction raise problematic issues. First, the proof seems to rest heavily on the assumption that there is a finite number of types. This seems restrictive, particularly since the allocation space is so large, namely the set of all simple lotteries over some abstract deterministic outcome space. This finiteness permits the authors to use finite mechanisms, which seem to be needed for the kind of construction they use. Second, the mechanisms are not balanced out of equilibrium. We know from other work in mechanism design that relaxing budget balancing, even out of equilibrium, can expand the set of implementable allocations. Third, the kind of games that they construct and to which the logic of iterated elimination of weakly dominant strategies is applied, are essentially the same kind of situations where game theorists (and replicable experiments) have shown that this iterative elimination procedure does not predict behavior very well.[50] Despite these potential problems, this line of work nicely complements earlier findings by Palfrey and Srivastava [1989b, 1991a], Abreu and Sen [1990], and Moore and Repullo [1988], that refinements of equilibrium lead to an ability to implement nearly any incentive compatible social choice function in many environments. Furthermore, the kinds of mechanisms used in the Abreu–Matsushima constructive proofs offer an alternative to the more traditional constructions in implementation theory that date back to the work of Maskin [1977].

ACKNOWLEDGEMENT

The authors would like to express their gratitude to Andrew Postlewaite and Matthew Jackson for reading the manuscript.

REFERENCES

Abreu, D. and H. Matsushima. (1990a). Virtual Implementation in Iteratively Undominated Strategies: Incomplete Information, mimeo, Princeton University.
Abreu, D. and H. Matsushima. (1990b). Exact Implementation, mimeo, Princeton University.
Abreu, D. and H. Matsushima. (1992). Virtual Implementation in Iteratively Undominated Strategies: Complete Information, *Econometrica*, **60**, 993–1008.
Abreu, D. and A. Sen. (1990). Subgame Perfect Implementation: A Necessary and Almost Sufficient Condition, *Journal of Economic Theory*, **50**, 285–299.
Abreu, D. and A. Sen. (1991). Virtual Implementation in Nash Equilibrium, *Econometrica*, **59**, 997–1021.
Aghion, P., M. Dewatripont and P. Rey. (1990). On Renegotiation Design, *European Economic Review*, **34**, 322–329.
Arrow, K. (1977). The Property Rights Doctrine and Demand Revelation under Incomplete Information, IMSSS Technical Report # 243, Stanford University.
Austen-Smith, D. and J. Banks. (1991). Monotonicity in Electoral Systems, *American Political Science Review*, in press.
Banks, J. and R. Calvert. (1992). A Battle of the Sexes Game with Incomplete Information, *Games and Economic Behaviour*, **4**, 347–72.
Baron, D. and D. Besanko. (1987). Commitment and Fairness in a Continuing Relationship, *Review of Economic Studies*, **54**, 285–299.

---

[50] This argument is made very clearly in Glazer and Rosenthal [1990].

Blume, L. and D. Easley. (1983). Implementation of Rational Expectations Equilibrium with Strategic Behavior, manuscript, Cornell University.

Blume, L. and D. Easley. (1990). Implementation of Walsrian Expectations Equilibria, *Journal of Economic Theory*, **51**, 207-227.

Chakravorti, B. (1992). Efficiency and Mechanisms with no Regret *International Economic Review*, **33**, 45-60.

Chakravorti, B. (1989). Sequential Rationality, Implementation, and Communication in Games, mimeo, University of Illinois.

Chatterjee, K. and W. Samuelson. (1983). Bargaining under Incomplete Information, *Operations Research*, **31**, 835-851.

Cramton, P. R., R. Gibbons, and P. Klemperer. (1987). Dissolving a Partnership Efficiently, *Econometrica*, **55**, 615-632.

Cramton, P. and T. Palfrey. (1989). Ratifiable Mechanisms: Learning from Disagreement, Working Paper # 731, California Institute of Technology.

Cramton, P. and T. Palfrey. (1990). Cartel Enforcement with Uncertainty about Costs, *International Economic Review*, **31**, 17-47.

Crawford, V. (1979). A Procedure for Generating Pareto-Efficient Egalitarian Equivalent Allocations, *Econometrica*, **47**, 49-60.

Crawford, V. (1985). Efficient and Durable Decision Rules: A Reformulation, *Econometrica*, **53**, 817-836.

Crawford, V. and J. Sobel. (1982). Strategic Information Transmission, *Econometrica*, **50**, 1431-1452.

d'Aspremont, C. and L-A. Gerard-Varet. (1979). Incentives and Incomplete Information, *Journal of Public Economics*, **11**, 25-1145.

Danilov, V. (1992). Implementation via Nash Equilibrium, *Econometrica*, **60**, 43-56.

Dasgupta, P., P. Hammond, and E. Maskin. (1979). The Implementation of Social Choice Rules: Some General Results on Incentive Compatibility, *Review of Economic Studies*, **46**, 185-216.

Demski, J. and D. Sappington. (1984). Optimal Incentive Contracts with Multiple Agents, *Journal of Economic Theory*, **33**, 152-171.

Dewatripont, M. (1989). Renegotiation and Information Revelation Over time: The Case of Optimal Labor Contracts, *Quarterly Journal of Economics*, **104**, 589-620.

Dutta, B. and A. Sen. (1991). A Necessary and Sufficient Condition for Two-Person Nash Implementation, *Review of Economic Studies*, **58**, 121-128.

Dutta, B. and A. Sen. Implementation Under Strong Equilibrium: A Complete Characterization, Indian Statistical Institute, mimeo.

Farquharson, R. (1957/1969). *Theory of Voting*. New Haven: Yale University Press.

Farrell, J. (1983). Communication in Games I: Mechanism Design Without a Mediator, mimeo, Massachusetts Institute of Technology.

Farrell, J. Meaning and Credibility in Cheap-Talk Games, forthcoming in *Mathematical Models in Economics* (ed. M. Demster), Oxford University Press.

Farrell, J. and R. Gibbons. (1989). Cheap Talk Can Matter in Bargaining, *Journal of Economic Theory*, **48**, 221-237.

Farrell, J. and G. Saloner. (1985). Standardization, Compatibility, and Innovation, *Rand Journal of Economics*, **16**, 70-83.

Fishburn, P. (1973). *The Theory of Social Choice*, Princeton: Princeton University Press.

Forges, F. (1986). An Approach to Communication Equilibria, *Econometrica*, **54**, 1375-1385.

Forges, F. (1988). Universal Mechanisms, GREQE Working Paper # 8810, University of Aix-Marseille.

Fudenberg, D. and J. Tirole. (1990). Moral Hazard and Renegotiation in Agency Contracts, *Econometrica*, **58**, 1279-1319.

Gibbard, A. (1973). Manipulation of Voting Schemes: A General Result, *Econometrica*, **41**, 587-601.

Glazer, J. and C-T. Ma. (1989). Efficient Allocation of a 'Prize' — King Solomon's Dilemma, *Games and Economic Behaviour*, **1**, 223-233.

Glazer, J. and R. Rosenthal. (1990). A Note on the Abreu-Matsushima Mechanism, mimeo, Boston University.

Green, J. and J-J. Laffont. (1977). Characterization of Satisfactory Mechanisms for the Revelation of Preferences for Public Goods, *Econometrica*, **45**, 427-438.

Green, J. and J-J. Laffont. (1979). *Incentives in Public Decision Making*, Amsterdam: North Holland.

Green, J. and J-J. Laffont. (1987a). Posterior Implementation in a Two-Period Decision Problem, *Econometrica*, **55**, 69-94.

Green, J. and J-J. Laffont. (1987b). Renegotiation and the Form of Efficient Contracts, HIER DP no. 1338.

Groves, T. (1982). On Theories of Incentive Compatible Choice with Compensation, in *Advances in Economic Theory* (ed. W. Hildenbrand) Cambridge University Press.

Groves, T. and J. Ledyard. (1977). Optimal Allocation of Public Goods: A Solution to the 'Free Rider' Problem, *Econometrica*, **45**, 783-809.

Groves, T. and J. Ledyard. (1987). Incentive Compatibility since 1972, in T. Groves, R. Radner, and S. Reiter (eds.), *Information, Incentives, and Economic Mechanisms: Essays in Honor of Leonid Hurwicz*. Minneapolis: University of Minnesota Press, pp. 48-111.

Guth, W. and M. Hellwig. (1986). The Private Supply of a Public Good. *Journal of Economics*, **5**, 121-59.

Harris, M. and A. Raviv. (1981). Allocation Mechanisms and the Design of Auctions, *Econometrica*, **49**, 1477-1499.

Harris, M. and R. Townsend. (1981). Resource Allocation with Asymmetric Information, *Econometrica*, **49**, 33-64.

Harsanyi, J. (1967, 1968). Games with Incomplete Information Played by Bayesian Players, *Management Science*, **14**, 159-182, 320-334, 486-502.

Hart, O. and J. Tirole. (1987). Contract Renegotiation and Coasian Dynamics, Review of Economic Studies, **55**, 509-540.

Herrero, M. and S. Srivastava. (1992). Implementation via Backward Induction, *Journal of Economic Theory*, **56**, 70-88.

Holmstrom, B. and R. Myerson. (1983). Efficient and Durable Decision Rules with Incomplete Information, *Econometrica*, **51**, 1799-1819.

Hurwicz, L. (1972). On Informationally Decentralized Systems, in *Decision and Organization (Volume in Honor of J. Marschzk)*, (eds. R. Radner and C. B. McGuire), pp. 197-336. Amsterdam: North Holland.

Hurwicz, L. (1979). Outcome Functions Yielding Walrasian and Lindahl Allocations at Nash Equilibrium Points, *Review of Economic Studies*, **46**, 217-225.

Hurwicz, L., E. Maskin, and A. Postlewaite. (1980). Feasible Implementation of Social Choice Correspondences by Nash Equilibria, mimeo, University of Minnesota.

Hurwicz, L., D. Schmeidler, and H. Sonnenschein (Eds.). (1985). *Social Goals and Social Organizations: Essays in Honor of Elisha Pazner*, Cambridge University Press, Cambridge.

Jackson, M. (1991). Bayesian Implementation, *Econometrica*, **59**, 461-477.

Jackson, M. (1989). Implementation in Undominated Strategies: A Look at Bounded Mechanisms, *Review of Economic Studies*, in press.

Jackson, M. and H. Moulin. (1990). Implementing a Public Project and Distributing Its Cost, *Journal of Economic Theory*, **57**, 125-40.

Jackson, M., T. Palfrey, and S. Srivastava. (1990). Undominated Nash Implementation in Bounded Mechanisms, *Games and Economic Behaviour*, in press.

Kihlstrom, R. E. and X. Vives. (1987). Collusion by Asymmetrically Informed Firms, Working Paper, University·of Pennsylvania.

Kohlberg, E. and J. F. Mertens. (1986). On the Strategic Stability of Equilibrium, *Econometrica*, **54**, 1003-1038.

Kydland, F. and E. Prescott. (1977). Rules Rather than Discretion: The Inconsistency of Optimal Plans, *Journal of Political Economy*, **85**, 473-491.

Laffont, J. and E. Maskin. (1982). The Theory of Incentives: An Overview, in *Advances in Economic Theory: Invited Papers for the Fourth World Congress of the Econometric Society at Aix-en Provence, September 1980*, (ed. W. Hildenbrand), Cambridge University Press, Cambridge, pp. 31-94.

Laffont, J. and J. Tirole. (1987). Comparative Statics of the Optimal Dynamic Incentive Contract, *European Economic Review*, **31**, 901-926.

Laffont, J.-J. and J. Tirole. (1988). The Dynamics of Incentive Contracts, *Econometrica*, **56**, 1153-1175.

Laffont, J. and J. Tirole. (1990). Adverse Selection and Renegotiation in Procurement, *Review of Economic Studies*, **57**, 597-625.

Ledyard, J. (1986). The Scope of the Hypothesis of Bayesian Equilibrium, *Journal of Economic Theory*, **39**, 59-82.

Ledyard, J. and T. Palfrey. (1989). Interim Efficient Public Good Provision and Cost Allocation with Limited Side Payments, Working Paper # 717, California Institute of Technology.

Legros, P. (1990). Strongly Durable Allocations, CAE Working Paper # 90-05, Cornell University.

Leininger, W. P., Linhart, and R. Radner. (1989). Equilibria of the Sealed Bid Mechanism for Bargaining with Incomplete Information, *Journal of Economic Theory*, **48**, 63-106.

Ma, C., J. Moore, and S. Turnbull. (1988). Stopping Agents from Cheating, *Journal of Economic Theory*, **46**, 355-372.

Mailath, G. and A. Postlewaite. (1990). Asymmetric Information Bargaining Problems with Many Agents, *Review of Economic Studies*, **57**, 351-367.

Maskin, E. (1977). Nash Equilibrium and Welfare Optimality, mimeo.

Maskin, E. (1985). The Theory of Implementation in Nash Equilibrium: A survey, in L. Hurwicz, D. Schmeidler, and H. Sonnenschein (eds.), *Social Goals and Social Organization: Essays in Memory of Elisha Pazner*. Cambridge: Cambridge University Press, pp. 173-204.

Maskin, E. and J. Moore. (1989). Implementation with Renegotiation, mimeo.

Maskin, E. and J. Riley. (1984). Optimal Auctions with Risk Averse Buyers, *Econometrica*, **52**, 1473-1518.

Maskin, E. and J. Tirole. (1990). The Principal-Agent Relationship with an Informed Principal, I: Private Values, *Econometrica*, **58**, 379-410.

Maskin, E. and J. Tirole. (1992). The Principal-Agent Relationship with an Informed Principal, II: Common Values, *Econometrica*, **60**, 1-42.

Matsushima, H. (1988). A New Approach to the Implementation Problem, *Journal of Economic Theory*, **45**, 128-144.

Matsushima, H. (1990a). Unique Bayesian Implementation with Budget Balancing, manuscript, Stanford University.

Matsushima, H. (1990b). Characterization of Full Bayesian Implementation, manuscript, Stanford University.

Matthews, S. (1989). Veto Threats: Rhetoric in a Bargaining Game, *Quarterly Journal of Economics*, **104**, 347-369.

Matthews, S. M., Okuno-Fujiwara, and A. Postlewaite, (1991). Refining cheap-talk Equilibria, *Journal of Economic Theory*, **55**, 247-73.

Matthews, S. and A. Postlewaite. (1989). Pre-play Communication in Two-Person Sealed-Bid Double Auctions, *Journal of Economic Theory*, **48**, 238-263.

May, K. O. (1952). A Set of Independent Necessary and Sufficient Conditions for Simple Majority Rule, *Econometrica*, **20**, 680-684.

McKelvey, R. and R. Niemi. (1978). A Multistage Game Representation of Sophisticated Voting for Binary Procedures, *Journal of Economic Theory*, **81**, 1-22.

Miller, N. (1977). Graph-Theoretic Approaches to the Theory of Voting, *American Journal of Political Science*, **21**, 769-803.

Mookherjee, D. and S. Reichelstein. (1989). Dominant Strategy Implementation of Bayesian Incentive Compatible Allocation Rules, manuscript, Stanford University.

Mookherjee, D. and S. Reichelstein. (1990a). Implementation Via Augmented Revelation Mechanisms, *Review of Economic Studies*, **57**, 453-475.

Mookherjee, D. and S. Reichelstein. (1990b). The Revelation Approach to Nash Implementation, mimeo, Indian Statistical Institute.

Moore, J. and R. Repullo. (1988). Subgame Perfect Implementation, *Econometrica*, **56**, 1191-1220.

Moore, J. and R. Repullo. (1990). Nash Implementation: A Full Characterization, *Econometrica*, **58**, 1083-1100.

Moulin, H. (1979). Dominance Solvable Voting Schemes, *Econometrica*, **47**, 1337-1352.

Moulin, H. (1983). *The Strategy of Social Choice*, North Holland, New York.

Mount, K. and S. Reiter. (1974). The Informational Size of Message Spaces, *Journal of Economic Theory*, **8**, 161-192.

Muller, E. and M. Satterthwaite. (1985). Strategy-proofness: The Existence of Dominant Strategy Mechanisms, in L. Hurwicz, D. Schmeidler and H. Sonnenschein (Eds.), *Social Goals and Social Organization: Essays in Memory of Elisha Pazner*. Cambridge: Cambridge University Press, pp. 131-171.

Myerson, R. (1979). Incentive Compatibility and the Bargaining Problem, *Econometrica*, **47**, 61-74.

Myerson, R. (1981). Optimal Auction Design, *Mathematics of Operations Research*, **6**, 58-73.

Myerson, R. (1983). Mechanism Design by an Informed Principal, *Econometrica*, **1**, 1767-1798.

Myerson, R. (1985). Bayesian Equilibrium and Incentive Compatibility: An Introduction, in L. Hurwicz, D. Schmeidler, and H. Sonnenschein (Eds.), *Social Goals and Social Organization: Essays in Memory of Elisha Pazner*. Cambridge: Cambridge University Press, pp. 229-259.

Myerson, R. (1986). Multistage Games With Communication, *Econometrica*, **54**, 323-358.

Myerson, R. and M. Satterthwaite. (1983). Efficient Mechanisms for Bilateral Trading, *Journal of Economic Theory*, **29**, 265-281.

Ordeshook, P. and T. Palfrey. (1988). Agendas, Strategic Voting, and Signalling with Incomplete Information, *American Journal of Political Science*, **32**, 441-466.

Palfrey, T. (1990). Implementation in Bayesian Equilibrium: The Multiple Equilibrium Problem in Mechanism Design, in J.-J. Laffont (Ed.) *Advances in Economic Theory*, Cambridge: Cambridge University Press, in press.

Palfrey, T. and H. Rosenthal. (1991). Testing for Effects of Cheap Talk in a Public Goods Game with Private Information, *Games and Economic Behavior*, **3**, 183-220.

Palfrey, T., and S. Srivastava. (1986). Private Information in Large Economies, *Journal of Economic Theory*, **39**, 34–58.

Palfrey, T. and S. Srivastava. (1987). On Bayesian Implementable Allocations, *Review of Economic Studies*, **54**, 193–208.

Palfrey, T. and S. Srivastava. (1989a). Implementation with Incomplete Information in Exchange Economies, *Econometrica*, **57**, 115–134.

Palfrey, T. and S. Srivastava. (1989b). Mechanism Design with Incomplete Information: A Solution to the Implementation Problem, *Journal of Political Economy*, **97**, 668–691.

Palfrey, T. and S. Srivastava. (1991a). Nash Implemenation Using Undominated Strategies, *Econometrica*, **59**, 479–501.

Palfrey, T. and S. Srivastava. (1991b). Efficient Trading Mechanisms with Preplay Communication, *Journal of Economic theory*, **55**, 17–40.

Postlewaite, A. (1985). Implementation via Nash Equilibria in Economic Environments, in L. Hurwicz, D. Schmeidler and H. Sonnenschein (Eds.), *Social Goals and Social Organization: Essays in Memory of Elisha Pazner*. Cambridge: Cambridge University Press, pp. 205–228.

Postlewaite, A. and D. Schmeidler. (1986). Implementation in Differential Information Economies, *Journal of Economic Theory*, **39**, 14–33.

Postlewaite, A. and D. Schmeidler. (1987). Differential Information and Strategic Behavior in Economic Environments: A General Equilibrium Approach. In T. Groves, R. Radner and S. Reiter (Eds.), *Information, Incentives and Economic Mechanisms – Essays in Honor of Leonid Hurwicz*. University of Minnesota Press.

Postlewaite, A. and D. Wettstein. (1989). Feasible and Continuous Implementation, *Review of Economic Studies*, **56**, 603–611.

Radner, R. (1979). Rational Expectations Equilibrium: Generic Existence and the Information Revealed by Prices, *Econometrica*, **47**, 655–678.

Radner, R. and A. Schotter. (1989). The Sealed Bid Mechanism: An Experimental Study, *Journal of Economic Theory*, **48**, 179–220.

Rajan, M. (1989). Cost Allocation in Multi-Agent Settings, Mimeo, Carnegie Mellon University.

Repullo, R. (1986). On the Revelation Principle with Complete and Incomplete Information, in *Economic Organizations as Games*, Oxford: Oxford University Press, pp. 179–795.

Repullo, R. (1987). A Simple Proof of Maskin's Theorem on Nash Implementation, *Social Choice and Welfare*, **4**, 39–41.

Riley, J. and R. Zeckhauser. (1983). Optimal Selling Strategies: When to Haggle, When to Hold Firm, *Quarterly Journal of Economics*, **98**, 267–287.

Rob, R. (1988). Pollution Claim Settlement Under Private Information, *Journal of Economic Theory*, **47**, 307–333.

Roberts, K. (1983). Self-Agreed Cartel Rules, IMSSS Working Paper, Stanford University.

Roberts, K. (1985). Cartel Behavior and Adverse Selection, *Journal of Industrial Economics*, 33–45.

Rubinstein, A. and A. Wolinsky. (1991). Renegotiation-proof Implementation and Time Preferences, *American Economic Review*, in press.

Saijo, T. (1988). Strategy Space Reduction in Maskin's Theorem: Sufficient Conditions for Nash Implementation, *Econometrica*, **56**, 693–700.

Satterthwaite, M. (1975). Strategy-Proofness and Arrow's Conditions: Existence and Correspondence Theorems for Voting Procedures and Social Welfare Functions, *Journal of Economic Theory*, **10**, 187–217.

Satterthwaite, M. and S. Williams. (1989). Bilateral Trade with the Sealed Bid k-Double Auction: Existence and Efficiency, *Journal of Economic Theory*, **48**, 107-133.

Schmeidler, D. (1980). Walrasian Analysis via Strategic Outcome Functions, *Econometrica*, **48**, 1585-1594.

Simon, L. and W. Zame. (1990). Discontinuous Games and Endogenous Sharing Rules, *Econometrica*, **50**, 861-872.

Sjostrom, T. (1990a). On the Necessary and Sufficient Conditions for Nash Implementation, mimeo, University of Rochester.

Sjostrom, T. (1990b). Implementation in Undominated Nash Equilibrium without Integer Games, mimeo, University of Rochester.

Stokey, N. (1979). Intertemporal Price Discrimination. *Quarterly Journal of Economics*, **93**, 355-371.

Thomson, W. (1979). Maximin Strategies and Elicitation of Preferences, in J-J. Laffont (Ed.), *Aggregation and Revelation of Preferences*. Amsterdam: North Holland, pp. 245-268.

Vickrey, W. (1961). Counterspeculation, Auctions, and Competitive Sealed Tenders, *Journal of Finance*, **16**, 1-17.

Walker, M. (1978). A Note on the Characterization of Mechanisms for the Revelation of Preferences, *Econometrica*, **46**, 147-152.

Wettstein, D. (1986). Implementation Theory in Economies with Incomplete Information, Working Paper # 26-86, Foerder Institute for Economic Research, Tel Aviv University.

Williams, S. (1984). Sufficient Conditions for Nash Implementation, mimeo, University of Minnesota.

Williams, S. (1986). Realization and Nash Implementation: Two Aspects of Mechanism Design, *Econometrica*, **54**, 139-151.

Williams, S. (1987). Efficient Performance in Two Agent Bargaining, *Journal of Economic Theory*, **41**, 154-172.

Wilson, R. (1985). Incentive Efficiency of Double Auctions, *Econometrica*, **53**, 1101-1115.

Yamato, T. (1990a). Sufficient Conditions for Implementation in Nash Equilibrium: Strong Monotonicity and Weak Unanimity, mimeo, University of Rochester.

Yamato, T. (1990b). Double Implementation in Nash and Undominated Nash Equilibria, University of Rochester, mimeo.

Zou, L. (1990). The Revelation Principle, Multiple Equilibria, and Communication Regimes, mimeo, Limburg University.

Index

FUNDAMENTALS OF PURE AND APPLIED ECONOMICS

SECTIONS AND EDITORS

BALANCE OF PAYMENTS AND INTERNATIONAL FINANCE
W. Branson, Princeton University
DISTRIBUTION
A. Atkinson, London School of Economics
ECONOMIC DEVELOPMENT STUDIES
S. Chakravarty, Delhi School of Economics
ECONOMIC HISTORY
P. David, Stanford University, and M. Lévy-Leboyer, Université
Paris X
ECONOMIC SYSTEMS
J.M. Montias, Yale University
ECONOMICS OF HEALTH, EDUCATION, POVERTY AND
CRIME
V. Fuchs, Stanford University
ECONOMICS OF THE HOUSEHOLD AND INDIVIDUAL
BEHAVIOR
J. Muellbauer, University of Oxford
ECONOMICS OF TECHNOLOGICAL CHANGE
F.M. Scherer, Harvard University
EVOLUTION OF ECONOMIC STRUCTURES, LONG-TERM
MODELS, PLANNING POLICY, INTERNATIONAL ECONOMIC
STRUCTURES
W. Michalski, O.E.C.D., Paris
EXPERIMENTAL ECONOMICS
C. Plott, California Institute of Technology
GOVERNMENT OWNERSHIP AND REGULATION OF
ECONOMIC ACTIVITY
E. Bailey, Carnegie-Mellon University, USA
INTERNATIONAL ECONOMIC ISSUES
B. Balassa, The World Bank
INTERNATIONAL TRADE
M. Kemp, University of New South Wales
LABOR AND ECONOMICS
F. Welch, University of California, Los Angeles, and J. Smith,
The Rand Corporation
MACROECONOMIC THEORY
J. Grandmont, CEPREMAP, Paris

MARXIAN ECONOMICS
J. Roemer, University of California, Davis
NATURAL RESOURCES AND ENVIRONMENTAL ECONOMICS
C. Henry, Ecole Polytechnique, Paris
ORGANIZATION THEORY AND ALLOCATION PROCESSES
A. Postlewaite, University of Pennsylvania
POLITICAL SCIENCE AND ECONOMICS
J. Ferejohn, Stanford University
PROGRAMMING METHODS IN ECONOMICS
M. Balinski, Ecole Polytechnique, Paris
PUBLIC EXPENDITURES
P. Dasgupta, University of Cambridge
REGIONAL AND URBAN ECONOMICS
R. Arnott, Boston College, Massachusetts
SOCIAL CHOICE THEORY
A. Sen, Harvard University
STOCHASTIC METHODS IN ECONOMIC ANALYSIS
Editor to be announced
TAXES
R. Guesnerie, Ecole des Hautes Etudes en Sciences Sociales,
Paris
THEORY OF THE FIRM AND INDUSTRIAL ORGANIZATION
A. Jacquemin, Université Catholique de Louvain

FUNDAMENTALS OF PURE AND APPLIED ECONOMICS

PUBLISHED TITLES

ISSN: 0191-1708